A Technical Guide to Mathematical Finance

A Technical Guide to Mathematical Finance covers those foundational mathematical topics most important to an aspiring or professional quant. The text goes beyond a simple recitation of methods and aims to impart a genuine understanding of the fundamental concepts underpinning most of the techniques and tools routinely used by those working in quantitative finance.

Features:

- Suitable for professional quants and graduate students in finance and mathematical/quantitative finance.
- "Concept Refreshers" used throughout to provide pithy summaries of ancillary topics.
- Step-by-step detail for formal proofs and mathematical descriptions.

Derek Zweig is the Chief Executive Officer and co-founder of Value Analytics, a data and analytics firm specializing in equity markets. Prior to founding Value Analytics, he worked as a capital markets risk analyst at a large regional bank, where he specialized in market and counterparty risk. Along with his experience in risk, Derek spent much of his career valuing business interests and intangible assets of private and public companies. He is a member of the CFA Institute and the Global Association of Risk Professionals. He has a graduate certificate in Financial Engineering from Columbia University, an M.S. in Applied Economics from Johns Hopkins University, and a B.S. in Finance from the Ohio State University. Derek regularly tutors children and young adults in mathematics, and enjoys rock climbing, pickup basketball and volleyball, and spending endless hours clowning around with his daughter.

A Technical Guide to Mathematical Finance

Derek Zweig

CRC Press is an imprint of the
Taylor & Francis Group, an **informa** business
A CHAPMAN & HALL BOOK

Front cover image: Min C. Chiu/Sutterstock

First edition published 2024
by CRC Press
2385 NW Executive Center Drive, Suite 320, Boca Raton FL 33431

and by CRC Press
4 Park Square, Milton Park, Abingdon, Oxon, OX14 4RN

CRC Press is an imprint of Taylor & Francis Group, LLC

ISBN: 978-1-032-68596-0 (hbk)
ISBN: 978-1-032-68723-0 (pbk)
ISBN: 978-1-032-68765-0 (ebk)

DOI: 10.1201/9781032687650

Typeset in Minion
by codeMantra

This book is dedicated to my daughter, Vivian, who will always be my greatest accomplishment.

Contents

Table of Figures

Table of Concept Refreshers

Acknowledgments

Much of the notation, structure, and prose in this book followed from thoughtful comments I received on early drafts. At times feeling paralyzed by indecision, it's difficult to imagine finishing this book without the insights from friends, colleagues, and mentors. I must specifically thank Shao Mou (Nelson) Cheng, Roberto Núñez (University of Missouri), Timothy Sumner, and Ton Dieker (Columbia University). Special thanks also go to Callum Fraser, Mansi Kabra, and the CRC Press editing team. Without their help, this book would be far less coherent.

SECTION 1

Introduction

THIS BOOK EMERGED FROM a humble collection of notes I compiled covering topics in finance. These notes gradually evolved, transforming from simple bullet points into comprehensive narratives with detailed explanations and proofs. What prompted me to embark on this journey? My passion for finance became clear at a young age. I followed a conventional career path – I took business classes, read a fraction of the countless books I bought, enrolled in relevant certifications, etc. However, upon reflection, I sensed a deep void in my knowledge of the fundamentals. I had dutifully memorized a myriad of equations and models frequently found in finance literature, and I knew which equations to employ depending on the problem at hand. I was aware of assumptions and limitations, but merely as a series of bullet points.

While these were effective means to pass tests, they didn't facilitate true understanding. The further I got in my education, the more I recognized the need to return to Finance 101. I needed to re-teach myself all the basics, but with an emphasis on the origins of my equation inventory, rather than the equations themselves.

Such information was not easy to attain. Even textbooks, while marginally helpful, often skip steps in derivations or exclude derivations entirely with simply a reference to another text. I found myself chasing citations back to decades-old papers written in dense prose and confusing notation. It took years to build an understanding of mathematics to fill gaps in derivations that other authors found seemingly trivial. Nonetheless, I gradually succeeded at my task of unearthing (and understanding) the sources of the many basic concepts taught in finance classes today.

DOI: 10.1201/9781032687650-1

As time progressed, I maintained what I had learned in an increasingly robust set of personal notes. I referenced them often, shared with friends and colleagues, and sought feedback whenever possible. After some reflection, I thought that others might find these notes useful. After several more years of struggling to reformulate my notes into readable form, I landed on a manuscript with a unique purpose: explaining the relatively advanced work that feeds into introductory finance.

Is there a market for such knowledge? The paradox of writing a book that explains introductory finance through an advanced lens is not lost on me. Having already written and refined most of the content in this book for personal use, I won't be disappointed if I remain the sole reader. On the other hand, reading a book beyond one's abilities is almost obligatory for a young financier, and I hope readers take a little with them for the next text.

Writing a book on first principles requires arbitrary decisions regarding the level of detail. Just how basic should I get? I leaned more toward including detail rather than assuming a reader's knowledge, but still had to cease explanations somewhere. To accommodate the gray areas, I sprinkled brief concept refreshers throughout this book. These may easily be skipped by an experienced reader, though they may be valuable to those like me who forget much of the minutiae from past education.

I must also apologize in advance if the content seems random or sporadic. Significant effort was put into structuring the book, but it likely won't align with any textbook taught in classes. My career, as is true of anyone, has been unique, and the content explored in this book reflects that. Accordingly, none of the sections provide a comprehensive treatment of the topics at hand. Rather, the content is carefully selected to provide technical context for material that is already prominent in finance courses.

The notation used in this book is meant to facilitate the reader's understanding rather than maximize rigor. While much of this book may appear technical at a glance, the math does not exceed undergraduate material. After all, this author is certainly no mathematician. This is part of the motivation behind providing more, rather than less, formulaic detail. The whole point is not to skip steps, or to skip as few as possible, so that the relative "layman" might follow along more easily. This has the unfortunate side effect of disguising the book as an overly technical manual to the bystander leafing through pages.

If one were to trace the topics in this book to their origins, they would find that most stem from papers or textbooks written in the mid-20th century. As it takes years and sometimes decades for new approaches to be adequately reviewed and tested, many of the topics in this book still make up the core undergraduate and graduate material in academic finance. With that said, one could write an entire book on more recent extensions of any single topic discussed. This is particularly true for Sections 4–6, which cover time series processes, derivative pricing, and econometric modeling.

Readers will notice a consistent theme throughout this book: the importance of *time*. Time value of money forms the bedrock of finance, upon which entire disciplines are built. The concept of time value was arguably the most important take-away from my undergraduate degree in finance. Understanding the role of time in corporate and quantitative finance is crucial for grasping more advanced exercises.

Recognition of time value of money dates back to at least 500 A.D., showing up in the Talmud, the governing text of Rabbinic Judaism. The Talmud tells of a case in which a dispute over the maturity date of a loan required damages to be paid. The damages were based on the difference in the value of a loan under varying terms. While vague, the story suggests an understanding of how value might be a function of time.

The concept is said to have been codified by Martin de Azpilcueta, also known as Doctor Navarrus of the Scholastics,[1] in his personal notes posthumously published by his nephew.[2] Navarrus was known for examining usury laws and the factors causing the exchange value of money to change over time. On Navarrus analyzing the nature of certain financial arrangements, we note the following quotation:

> Indeed, Navarrus [Azpilcueta] himself, treating of the sale of debts at a discount, concedes that such sales are lawful, 'both because a claim on something is worth less than the thing itself, and because it is plain that that which is not usable for a year is less valuable than something of the same quality which is usable at once'. [3]

Modern notions of time value of money are more concrete with respect to opportunity cost. Such a framing might have been impractical in the time of the Scholastics, as usury laws were strict, jurisdiction-specific, and changing over time. Opportunity cost was, and is, legalistic by nature. The mechanisms that drive time value necessarily require freedom to pursue alternative uses. This will become clearer in our description of time value of money in Section 2.1.

A proper definition of the time value of money facilitates asset pricing through *discounted cash flow* (DCF) modeling. DCF models assume that an asset's value stems from the present value of all future cash flows accruing to the owner of the asset. While vague notions of DCF analysis date back to ancient Babylon, such models gained prominence in the early 19th-century British coal industry for estimating the accounting value of collieries.[4] Discount rates at this time were viewed as a risk-free rate plus a risky spread, and sensitivity analysis was often used to determine the impact of different risky spreads on the colliery value.

The modern approach to DCF analysis was codified by Irving Fisher in his famous text *The Theory of Interest*,[5] published in 1930. While application of this approach has evolved overtime, the core concept is unchanged. Discounting (and compounding) allows for intertemporal comparison of both risk-free and risky cash flows. This provides the backbone of modern corporate finance and complements a host of methods in quantitative finance.

The remainder of this book is structured as follows. In Section 2, we discuss the importance of time and cover basic concepts in finance. In Section 3, we study the cost of capital and further develop its relationship with financial asset prices. Sections 4 and 5 contextualize financial asset prices when the cost of capital is random, and the randomness behaves in a structured way. Section 6 explores what we can learn when the cost of capital is a function of exogenous variables. Section 7 drops the structured behavior of randomness from Sections 4 and 5, allowing us to observe how the cost of capital impacts financial asset values in the face of true uncertainty. Sections 8 and 9 are a continuation of Section 3, expanding on our understanding of the information contained in the cost of capital.

1.1 NOTATION AND FORMATTING

The following list describes the notation used in this text. Much of the notation is standard but is defined in more detail for those unfamiliar.

Many topics are touched on in this book, each requiring an inventory of variables to represent objects in formulas. Capital letters, lowercase letters, and a series of other symbols are used to this end. Conventional definitions for variables are used when possible. Variables are generally selected to maximize reader comprehension, not to maximize rigor. It is possible that some variables will be recycled from one section to another. Variable definitions will always be included when the notation changes.

When dealing with random variables, it is common for texts to differentiate between a *random variable* X and a *realization* of the random variable, x. Loosely speaking, random variables take on values according to a frequency distribution. Consider a random variable X that takes the values 0 or 1. This means that the only possible outcomes are $x=0$ or $x=1$. The capitalization signifies the abstraction of the random variable, whereas the lowercase signifies an actual outcome. Introducing some confusion, it is also common for non-letter symbols to signify random variables and subscripts to differentiate between the abstract random variable and its realizations. While both of these notational practices are used in this text, rigid attachment to rigorous notation is not placed above reader comprehension. Deviations from standard notation will be clear from context. While random variables will be denoted with a capital letter or non-letter symbol, this does not mean all capital letters or non-letter symbols indicate a random variable. When a random variable is introduced, it will be identified as such.[6]

$E(X)$ represents the expected value of random variable X (i.e., the weighted average function $\sum_{i=1}^{n} x_i p(x_i)$ for a discrete random variable or $\int_{-\infty}^{\infty} xf(x)\,dx$ for a continuous random variable, where $p(x_i)$ is the probability of realization x_i and $f(x)$ is the density function of X). $\bar{X}$ and μ_X may also be used to express the expected value.

$\text{Var}(X)$ represents the variance of random variable X (i.e., the variance function $E(X-\bar{X})^2$). Note that

$$
\begin{aligned}
E\left[X-\bar{X}\right]^2 &= \sum_{i=1}^{n}\left(x_i-\bar{X}\right)^2 p(x_i) \\
&= \sum_{i=1}^{n}\left(x_i^2-2x_i\bar{X}+\bar{X}^2\right)p(x_i) \\
&= \sum_{i=1}^{n}x_i^2 p(x_i)-2\bar{X}\sum_{i=1}^{n}x_i p(x_i)+\bar{X}^2\sum_{i=1}^{n}p(x_i) \\
&= E\left(X^2\right)-2\bar{X}^2+\bar{X}^2 \\
&= E\left(X^2\right)-\bar{X}^2 \\
&= E\left(X^2\right)-E(X)^2
\end{aligned}
$$

since $\sum_{i=1}^{n} p(x_i)=1$ and $\sum_{i=1}^{n} p(x_i)x_i = \bar{X}$. σ^2 will also be used to express the variance.

$\frac{\Delta f}{\Delta x}$ represents the change in function f for a change in variable x.

$\frac{df}{dx} = \lim_{\Delta x \to 0} \frac{\Delta f}{\Delta x}$ represents how function f changes for an infinitesimally small change in the variable x. This is also known as the first derivative of function f with respect to variable x. $\frac{\partial f}{\partial x}$ and $f'(x)$ may also be used interchangeably to represent the first derivative of function f with respect to a change in x. In the latter representation, each additional apostrophe indicates the derivative of an additional order (i.e., $f''(x)$ is the second derivative of f, etc.).

$\in$ should be read as "is an element of" or "is in." If S represents the set $(1,2,3)$, and $s \in S$, then s must be equal to 1, 2, or 3.

$\sim$ should be read as "is distributed as." This is generally followed by a probability distribution assumption. For example, if some random variable X is normally distributed, we write $X \sim N(\mu,\sigma^2)$, where N indicates a normal distribution, μ is the mean, and σ^2 is the variance.

$\propto$ should be read as "is proportional to." The term "proportional to" simply means multiplication by some scalar. For example, if $y \propto x$, this just means that $y = kx$ where k is some constant.

$\{x_t\}_{t=1}^{T}$ represents a sequence of values of x indexed by the variable t. If $T = 3$, then $\{x_t\}_{t=1}^{T}$ is equivalent to the sequence $\{x_1, x_2, x_3\}$.

$\{x : x^2 = 9\}$ should be read as "the values of x such that $x^2 = 9$." More generally, a colon in this context should be read "such that."

A vector or matrix will be represented by a bolded variable. For example, $\boldsymbol{x} = [x_1,\ldots,x_n]$ is a vector with n elements. $\boldsymbol{x}^T = \begin{bmatrix} x_1 \\ \vdots \\ x_n \end{bmatrix}$ is the transpose of the vector.

Both e^x and $\exp(x)$ will be used interchangeably.

Color formatting is occasionally used to help the reader follow from equation to equation. There are three particular instances of this the reader should be aware of.

1 *Colored terms on the same line*. For example,

$$y = 2x^2 + 3x + 5x^4 + 7 - 3x$$

$$y = 2x^2 + 5x^4 + 7$$

Here, the two terms are colored red because they cancel each other out.

2 *Colored terms on consecutive lines*. For example,

$$y = (x+4)^2 - 6x + 9$$

$$y = x^2 + 8x + 16 - 6x + 9$$

Here, the two terms are colored blue on consecutive lines to represent the fact that $(x+4)^2$ was expanded into $x^2 + 8x + 16$ on the following line.

3 *Colored terms are used for emphasis*. For example,

$$y = \frac{df}{dx} + 2\frac{d^2 f}{dx^2} + 3x - 5$$

Here, the blue terms turn this function into a differential equation, while the red terms are simpler functional terms. Notice the blue and red coloring is not used to explain any reduction or simplification, but it is just used for emphasis. If coloring is used for emphasis only, it will be accompanied by explanatory text.

The majority of color formatting will follow these conventions. If color formatting is used in a different way, it will be specified at the time it is used.

Lastly, the reader will notice that some equations are numbered, while others are not. Numbering was used only for those equations that are referenced later in the book. If an equation is not referenced, it does not receive a number.

NOTES

1 The School of Salamanca represents an intellectual movement of Spanish theologians in the 16th and 17th centuries. They are known for writings on early political economy, commerce, and finance. The movement is generally recognized as promoting natural law in lieu of medieval notions of law.

2 *Consiliorum sive responsorum libri quinque, iuxta ordinem decretalium dispositi* published in 1590, four years after Navarrus' death.

3 Caranti (2020) citing Noonan (1957). There is an interesting debate as to the legitimacy of Navarrus' claim as the inventor of time value of money. The quote provided is the only evidence amongst his many works that suggest an understanding. Caranti (2020) notes several passages, particularly from de Azpilcueta (2004), that appear to contradict the idea that Navarrus understands time value. While Navarrus currently holds claim to this discovery in several sources, it is this author's opinion that Caranti is correct and Navarrus coincidentally describes time value without demonstrating an understanding of it.

4 Brackenborough et al. (2001). Note that cash flow projections in the British coal industry date back at least an additional 100 years, but lacked the discounting component. It is believed that high-profit prospects of coal mines from new mining technology in the late 18th century attracted interest from emerging national capital markets, increasing the importance of discounting to account for risk differences between projects. The use of discounting caught on slowly after perpetuation by big names in the colliery valuation industry.

5 Fisher (1930).

6 Random variables won't be introduced until Section 4.

SECTION 2

Basics

2.1 TIME VALUE OF MONEY

Shopping for the best value is easy for the average consumer. Say you're interested in purchasing a bicycle. You may search local or online retailers for options, making note of quality and aesthetic. Each bicycle will have a price that is comparable to the prices of other bicycles. In this context, the prices are *comparable* because payment for the chosen bicycle is made at a single point in time.

What if each bicycle had not only a different price but also required payment at a different point in time? As we will see, these prices are no longer comparable. In other words, value is not time-invariant. As one moves forward or backward through time, value changes in either a predictable or unpredictable way. In the former case, we refer to changes as deterministic. In the latter case, changes are random (i.e., *stochastic*).

Let's consider a simple example to demonstrate this claim. Say you are offered two options: option (A) you receive $100 today, and option (B) you receive $100 in one year. Should you be indifferent between options A and B?

Of course not. If you choose option A, you can invest the $100 in order to earn some return over the coming year. While savings rates may at times be quite low, to make the concept clear, let's assume banks are paying 5% interest per year (paid at the end of the year) on money deposited into a savings account. In such a case, we can state both options A and B in terms of value now (present value), and value in one year (future value).

DOI: 10.1201/9781032687650-2

In one year, by definition, option B is worth \$100. Option A, due to the ability to generate a 5% return, will earn an additional $\$100 \times 0.05 = \$5$ by the end of the year. This makes option A worth

$$\begin{aligned} &\$100 + (\$100 \times 0.05) \\ &= \$100[1 + 0.05] \\ &= \$105 \end{aligned}$$

in one year. So in one year, option A is worth \$105, while option B is worth \$100. Let's make sure we're interpreting what just happened correctly. We defined option A as having a present value of \$100 and discovered that the future value (in one year) is \$105. Generalizing the step colored red, we state the relationship to be

$$\text{Present value} \times (1 + \text{return}) = \text{Future value}$$

In similar fashion, we state this relationship in terms of present value:

$$\text{Present value} = \frac{\text{Future value}}{(1 + \text{return})}$$

With this in mind, let's make the same value comparison but for the present time. Again, by definition, the present value of option A is \$100. To find the present value of option B, we use what we just learned. We know the future value of option B is, again given by definition, \$100. And we know that we could earn 5% per year, paid at the end of the year if we deposited money into a savings account today. Plugging these values, we see that

$$\begin{aligned} \text{Present value of option B} &= \frac{\$100}{(1 + 0.05)} \\ &= \$95.24 \end{aligned}$$

Option B is worth only \$95.24 today. In other words, one would only need to invest \$95.24 today at the prevailing 5% interest rate to generate the \$100 payout that option B is offering in one year. In both cases, we see that option A is superior to option B, even though they are both the same offer separated only by time!

The 5% rate of return paid by the bank is referred to as the *opportunity cost of capital*. Opportunity cost is often taught in simpler terms (e.g., if you go to happy hour after work then you can't go to the gym). The opportunity cost of capital represents the most lucrative alternative use of capital (money).[1] You may say, "honestly, if I had $100 today I would just spend it on consumption, leaving me with nothing in one year." This is certainly a valid alternative use, but it is not the most lucrative. While consumption qualifies as one possible opportunity cost of choosing option B, it is not useful for defining the opportunity cost of capital. The opportunity cost of capital will be discussed in more detail in Section 3.1.

One last note on terminology. When a value is multiplied by one plus a rate, it is referred to as *compounding*. When a value is divided by one plus a rate, it is referred to as *discounting*. Discounting is conceptually just the reversal of compounding. While value compounds forward through time, it discounts backward through time. As we will see throughout this book, time value of money is a central topic. It underlies all of finance. It's important for readers to understand not only the math but also the intuition behind the math.

2.2 CONTINUOUS VS. DISCRETE COMPOUNDING

As we saw in Section 2.1, the value of a payment made or received changes as you move forward or backward through time. Notice that the 5% annual return to saving money at the bank was specified to be paid at the end of the year. While this makes calculations easy, returns might accrue at all sorts of frequencies. In the case of *compounding* interest, investors are allowed to earn interest on top of interest each time interest accrues.[2]

To understand the implications of compounding interest better, let's say we start with P dollars and we wish to find the future value V when earning annual interest rate r over T years. The most basic situation involves a *compounding period* of one year. The compounding period is the period of time over which interest is calculated and accrued. A compounding period of one year is referred to as annual compounding, wherein interest compounds only once over the course of the year (at the very end). Under annual compounding, the future value is simply:

$$V = P(1+r)^T \quad (1)$$

Reading this qualitatively, consider the situation in which one invests P for one year $(T=1)$ earning an annually compounding interest rate of 5% $(r=5\%=0.05)$. Then

$$V = P(1+0.05) = 1.05P$$

In other words, an investment today of P will be worth $1.05P$ in one year.

If interest compounds twice in one year, known as *semi-annual* compounding, this simply implies that interest is accrued twice during the year (once at six months and once at the end of the year). The value in one year is then

$$V = P\left(1+\frac{r}{2}\right)\left(1+\frac{r}{2}\right)$$

If r still equals 5%, then this implies that

$$\begin{aligned} V &= P\left(1+\frac{0.05}{2}\right)\left(1+\frac{0.05}{2}\right) \\ &= 1.0506P \end{aligned}$$

We are dividing r by two because the amount of interest earned each half-year is only $\frac{r}{2}$. Notice that we are left with a slightly higher multiple of P using semi-annual compounding (a multiple of 1.0506) relative to annual compounding (a multiple of 1.05). This is because the interest earned over the first six months is itself also earning interest over the second six months. As the compounding frequency increases, interest is accrued more often and builds upon itself more quickly.

Restating semi-annual compounding for any number of years, we have

$$V = P\left(1+\frac{r}{2}\right)^{2T}$$

Similarly, with monthly compounding

$$V = P\left(1+\frac{r}{12}\right)^{12T}$$

Generalizing, if the number of compounding periods per year is n, then

$$V = P\left(1+\frac{r}{n}\right)^{nT} \tag{2}$$

As the compounding frequency n approaches infinity, we have what is referred to as *continuous compounding*. Continuous compounding can be thought of as a compounding period so infinitesimally small that interest is earned and accrued an infinite number of times each year. The rate earned each infinitesimally small period of time is $\frac{r}{n}$ as $n \to \infty$, while the number of compounding periods per year is nT as $n \to \infty$. We then have

$$V = \lim_{n \to \infty} P\left(1 + \frac{r}{n}\right)^{nT}$$

It turns out that this can be reduced into a simpler expression. To start, we divide both sides by P:

$$\frac{V}{P} = \lim_{n \to \infty}\left(1 + \frac{r}{n}\right)^{nT}$$

The left-hand side now represents the ratio of the future value to the present value, and we have isolated the return to be earned by the investor on the right-hand side. Taking the natural log of both sides gives us

$$\ln\left(\frac{V}{P}\right) = \ln\left[\lim_{n \to \infty}\left(1 + \frac{r}{n}\right)^{nT}\right]$$

The natural log function is continuous for all positive numbers. As long as $\frac{r}{n} > -1$ (i.e., the interest rate applied over the compounding period is greater than -100%), the function inside the natural log will never be zero or negative. As a result, the natural log may be moved inside the limit function, giving us

$$\begin{aligned} \ln\left(\frac{V}{P}\right) &= \lim_{n \to \infty} \ln\left[\left(1 + \frac{r}{n}\right)^{nT}\right] \\ &= \lim_{n \to \infty} nT \ln\left[1 + \frac{r}{n}\right] \end{aligned} \tag{3}$$

To write the natural log function differently, we consider a Taylor series. A Taylor series equates a function to an infinite sum of terms, with each term being a derivative of the function of increasing order. Taylor series has the following form:

$$f(x) = \sum_{d=0}^{\infty} \frac{f^{(d)}(a)}{d!}(x - a)^d$$

where a is any convenient point in the function where the derivative may be taken repeatedly and the superscript (d) refers to the order of the derivative.

CONCEPT REFRESHER 2.1: TAYLOR EXPANSION

A full explanation of a Taylor expansion is beyond the scope of this book. With that said, for a basic intuition, consider the Taylor series written out:

$$f(x)=\sum_{d=0}^{\infty}\frac{f^{(d)}(a)}{d!}(x-a)^{d}=f(a)+\frac{f'(a)}{1!}(x-a)+\frac{f''(a)}{2!}(x-a)^{2}+\cdots$$

Notice the first term of the series is $f(a)$. In words, the Taylor series is first finding the value of the function at some convenient value a and then adding an adjustment to approximate the value of the function at x. One might write $f(x)=f(a)+\text{adjustment}$.

Those familiar with calculus should understand that the derivative $f'(a)$ provides an approximation of the sensitivity of $f(a)$ for a change in a. Loosely speaking, the adjustment terms are simply the sensitivities $f^{(d)}(a)$ of the function multiplied by the magnitude of the change $(x-a)$ and a scalar $\frac{1}{d!}$. Since $f'(a)$, being the first derivative, provides only a linear sensitivity, additional terms are added to capture nonlinearities in the function. As the number of derivatives approaches infinity (i.e., as $d\to\infty$), the Taylor expansion becomes an increasingly accurate approximation of $f(x)$.

Treating the natural log as the function f and using $x=1+\frac{r}{n}$ and $a=1$, the Taylor series of $\ln\left[1+\frac{r}{n}\right]$ is[3]

$$\begin{aligned}\ln\left[1+\frac{r}{n}\right]&=\ln(a)+\frac{a^{-1}}{1!}\left(1+\frac{r}{n}-a\right)^{1}\\&\quad+\frac{-a^{-2}}{2!}\left(1+\frac{r}{n}-a\right)^{2}+\frac{2a^{-3}}{3!}\left(1+\frac{r}{n}-a\right)^{3}+\cdots\\&=0+\frac{r}{n}-\frac{1}{2}\left(\frac{r}{n}\right)^{2}+\frac{2}{6}\left(\frac{r}{n}\right)^{3}+\cdots\\&=\frac{r}{n}-\frac{r^{2}}{2n^{2}}+\frac{r^{3}}{3n^{3}}-\frac{r^{4}}{4n^{4}}+\cdots\end{aligned}\tag{4}$$

Plugging Equation 4 into Equation 3, we get

$$\begin{aligned}
\ln\left(\frac{V}{P}\right) &= \lim_{n\to\infty} nT\ln\left[1+\frac{r}{n}\right] \\
&= \lim_{n\to\infty} nT\left\{\frac{r}{n}-\frac{r^2}{2n^2}+\frac{r^3}{3n^3}-\frac{r^4}{4n^4}+\cdots\right\} \\
&= \lim_{n\to\infty}\left\{\frac{nTr}{n}-\frac{nTr^2}{2n^2}+\frac{nTr^3}{3n^3}-\frac{nTr^4}{4n^4}+\cdots\right\} \\
&= \lim_{n\to\infty}\left\{rT-\frac{r^2T}{2n}+\frac{r^3T}{3n^2}-\frac{r^4T}{4n^3}+\cdots\right\}
\end{aligned}$$

Distributing the limit operator, we get

$$\ln\left(\frac{V}{P}\right) = \lim_{n\to\infty}(rT) - \lim_{n\to\infty}\left(\frac{r^2T}{2n}\right) + \lim_{n\to\infty}\left(\frac{r^3T}{3n^2}\right) - \lim_{n\to\infty}\left(\frac{r^4T}{4n^3}\right) + \cdots$$

Notice that other than the first term on the right-hand side, all terms approach zero as $n\to\infty$. Thus we are left with

$$\ln\left(\frac{V}{P}\right) = \lim_{n\to\infty}(rT)$$

Since n is no longer present on the right-hand side, the limit operator may be dropped completely, leaving

$$\ln\left(\frac{V}{P}\right) = rT$$

Reversing the natural log function gives us

$$\frac{V}{P} = e^{rT}$$

and multiplying through once more by P gives us our final form

$$V = Pe^{rT} \tag{5}$$

Here we have the future value when the interest rate is continuously compounded.

NOTES

1 The opportunity cost of capital more accurately represents the most lucrative alternative use of capital commensurate with an equal level of risk. More on this in Section 3.1.

2 When interest accrues, it is added to the balance of money saved. This makes the balance increase by the amount of the interest, implying that the next time interest is calculated it will be based on the original balance plus all previous accrued interest. This is how interest compounds and what it means to earn interest on interest. This contrasts with *simple interest*, wherein the interest payment is always based on the original balance (without adding previously accrued interest).

3 Note that $f'(a) = \frac{d}{da}\ln(a) = \frac{1}{a} = a^{-1}$. Similarly $f''(a) = -a^{-2}$ and $f'''(a) = 2a^{-3}$.

SECTION 3

Fixed Income

We now have the tools to find future value for all sorts of different compounding schemes. This includes annual compounding from Equation 1

$$V = P(1+r)^T$$

continuous compounding from Equation 5

$$V = Pe^{rT}$$

and any compounding frequency in between from Equation 2

$$V = P\left(1+\frac{r}{n}\right)^{nT}$$

As we did earlier, with just a bit of algebra, we can restate these relationships in terms of P, allowing us to calculate present value when given future value. In continuous compounding terms, we can rearrange Equation 5 to get

$$P = \frac{V}{e^{rT}}$$

$$= Ve^{-rt}$$

In annual compounding terms, we rearrange Equation 1 to get

$$P = \frac{V}{(1+r)^T}$$

DOI: 10.1201/9781032687650-3

This provides a clean transition into fixed-income securities. Bonds are one of the simplest securities because their payoff is generally deterministic. Bonds pay periodic coupons (which are essentially interest payments) and then return the par value, also known as the *principle*, at the end of some predetermined period of time. The simplest bond has a fixed rate and makes coupon payments at the end of each year (i.e., annually). Let's look at an easy example.

Say we have a bond with a par value of \$1,000 that pays a 5% coupon annually over the next three years. At 5% annually, the coupon payment will be $\$1{,}000 \times 0.05 = \$50$ at the end of each year. The value of this bond can be found as a simple extension of the present value formula just presented.

$$\text{Value of bond} = \frac{\$50}{1+r} + \frac{\$50}{(1+r)^2} + \frac{\$1{,}050}{(1+r)^3}$$

Notice that at the end of year three, both the coupon and the par value are paid. Viewing each cash flow independently, what we have here is a sequence of separate present value calculations, each over a different time period. If we assume an opportunity cost of capital of 3%, the value of the bond would equal

$$1{,}056.57 = \frac{\$50}{1.03} + \frac{\$50}{(1.03)^2} + \frac{\$1{,}050}{(1.03)^3}$$

Generalizing this formula, the value B of a fixed-rate, annual-pay bond that expires in n years with cash flow c_i in t_i years and opportunity cost of capital r is

$$B = \sum_{i=1}^{n} \frac{c_i}{(1+r)^{t_i}} \tag{6}$$

Once again, the last cash flow c_n includes the return of the bond's principle. Note that the opportunity cost of capital r, in the context of pricing a bond, is also referred to as the *yield to maturity* (YTM). With this formula, one has the tools to assess the value of any fixed-rate bond.

CONCEPT REFRESHER 3.1: YIELD TO MATURITY AND INTERNAL RATE OF RETURN

When pricing a coupon-paying bond, where each coupon is paid at a different point in time, it is necessary to discount each cash flow at a rate commensurate with the time until payment. We can simplify the pricing formula by using the YTM, conceptually the same as the *internal rate of return* (IRR) of the bond. This is best demonstrated using an example.

Consider a fixed-rate bond with a par value of \$100 and 5% annual coupon that matures in three years. Modeling the price B, we have

$$B = \frac{5}{1+r_1} + \frac{5}{(1+r_2)^2} + \frac{105}{(1+r_3)^3}$$

where r_t is the *zero rate*[1] used to discount the cash flow at time t. Valuing a bond this way requires knowledge of zero rates for every time t. While it would be convenient to just use a constant discount rate across terms, in reality, the term structure of interest rates is rarely or never flat such that all interest rates are the same for different time horizons. Say zero rates are as follows:

Time Horizon	Zero Rate
1 Year	0.04
2 Year	0.05
3 Year	0.06

The bond price B is then

$$97.50 = \frac{5}{1.04} + \frac{5}{(1.05)^2} + \frac{105}{(1.06)^3}$$

Calculating the sensitivity of B to all rates is generally not feasible, as one would have to account for every combination of changes in zero rates. Instead, we can replace the zero rates r_1, r_2, and r_3 with a single rate r that recovers this price. Since we know that the price of the bond is $B = 97.50$, we have

$$97.50 = \frac{5}{1+r} + \frac{5}{(1+r)^2} + \frac{105}{(1+r)^3}$$

With a bit of trial and error, it turns out the value $r \approx 0.0593$ makes this identity hold. $r = 0.0593$ is the YTM of this bond. YTM is comprised of three components: (1) the coupon rate, (2) bond price appreciation, and (3) reinvestment return. If the coupon rate is equal to the YTM, the bond will trade at its par value (i.e., no premium or discount).

When applied in project finance, the value of r that sets the net present value (NPV) of a project equal to \$0 is referred to as the IRR. Thus, YTM and IRR have the same interpretation but with different contexts.

3.1 OPPORTUNITY COST OF CAPITAL[2]

The opportunity cost of capital (denoted r throughout Section 2) plays an important role when allocating capital to competing uses, a process known as capital budgeting. It will be useful to define the concept more rigorously.

Capital budgeting grew in importance in the United States amidst the capital-rich post-World War II environment, as managers sought objective criteria for choosing one project over another. For a profit-seeking entity, the goal of capital budgeting is generally to maximize the value of its equity. The capital budgeting process then entails (1) estimating potential profitability across competing capital-consuming projects and (2) using these estimates to assess a minimum profitability threshold required to make an investment.

To give more context, let's define three rates of return:

- Risk-free rate $\left(r_f\right)$
- Marginal project return $\left(r_p\right)$
- Expected rate of profit $\left(r_k\right)$

The risk-free rate represents the return one receives without any risk of loss. In other words, there is a 100% probability of receiving a risk-free return. Some consider the risk-free rate to be a manifestation of time preference. According to this view, the risk-free rate reflects the equilibrium compensation the average person requires to forego present consumption in favor of future consumption. In practice, the risk-free rate is proxied by collateralized short-run borrowing rates or sovereign borrowing rates, both of which are impacted by policy decisions.

The marginal project return represents the return an entity expects to receive on the next project it pursues.[3] Projects generally require capital outlays up-front, with cash inflows occurring at future dates. The most common proxy for the marginal project return is the IRR. The IRR represents the implied opportunity cost of capital that equates the value of net future cash inflows to capital outlays made at the beginning of the project. Said differently, discounting at the IRR produces an NPV of zero (i.e., the discounted future inflows exactly equal the capital outlays).[4]

The expected rate of profit represents the rate of return that market participants would expect to receive for investing in a security of equal risk.

Consider the basic example of purchasing an AAA-rated bond. While this high rating suggests a low risk of loss, the risk of not receiving promised cash flows is not zero. To determine an appropriate remuneration for taking this risk, one might review a basket of alternative AAA-rated bonds. If the return of the bond is significantly lower than that of the basket, it would be foolish to purchase it, as one could receive a higher return for taking the same risk by purchasing a bond from the basket. By benchmarking expected rates of profit across securities, one implicitly determines the market's risk assessment of the securities themselves.[5]

As we will see, the opportunity cost of capital is effectively equivalent to the expected rate of profit r_k. Consider a very basic asset with a single cash flow c in some future time period. Using what we already know about time value, the value P of this asset today (at time $t = 0$) is

$$P_0 = \frac{c}{1+r} \tag{7}$$

where r is the opportunity cost of capital. Equivalently, this can be written as

$$P_0(1+r) = c \tag{8}$$

illustrating simply that r serves simultaneously as the opportunity cost of capital on the right-hand side of Equation 7 and the expected rate of profit on the left-hand side of Equation 8. It's just a matter of perspective. The same works if the asset has two future payments, in which case we have

$$P_0 = \frac{c}{1+r} + \frac{c}{(1+r)^2} \tag{9}$$

Moving forward one time-increment at a time gives us

$$P_0(1+r) = c + \frac{c}{1+r}$$

$$P_0(1+r)^2 = c(1+r) + c \tag{10}$$

As each period passes, the value P grows by $1+r$. The opportunity cost of capital is equivalent to the expected rate of profit. This treatment implies that each cash flow can be reinvested to earn the same rate of profit,[6] as demonstrated by $c(1+r)$ in Equation 10.

Since markets assign the expected rate of profit r_k as the return commensurate with the perceived level of risk, markets are essentially opining as to what the opportunity cost of capital for the entity *should* be. If the marginal project return r_p does not equal r_k, then the identity from Equations 7–10 will fail to hold. Investing in a project where $r_p > r_k$ implies the price P_0 should grow by $1 + r_p$, while the opportunity cost of capital used for discounting is only r_k. Since $1 + r_p > 1 + r_k$, investing in these types of projects will grow the equity of the entity. One might represent this mathematically by defining $c = P_0(1 + r_p)$ and rewriting Equation 7 as

$$P_0' = \frac{c}{1 + r_k}$$

$$= \frac{P_0(1 + r_p)}{1 + r_k}$$

Since $r_p > r_k$, $P_0' > P_0$. The value has grown. Instead, if $r_p < r_k$, investing in such projects will reduce the equity of the entity by the same logic.

It should now be clear that maximizing the value of equity necessitates that capital budgeters continue to invest in projects as long as $r_p > r_k$, stopping only when $r_p = r_k$. If one more project is invested in, such that $r_p < r_k$, value is reduced. If one project is left on the table where $r_p > r_k$, value is not as high as it could be. At $r_p = r_k$, the capital budgeter is indifferent between investing and not investing in a project. Stated differently, if a project's cash flows are discounted at the opportunity cost of capital and the present value of future cash inflows exceeds project outlays,[7] the capital budgeter should always pursue the project. Capital budgeting is discussed more in Section 8.1.

3.2 GORDON GROWTH MODEL[8]

We now understand that the expected rate of profit is equivalent to the opportunity cost of capital. We can combine these terms by referring to them as the *cost of capital*. Let's see what we can learn about the cost of capital by generalizing from an asset with just one or two cash flows, to an asset with infinite future cash flows (c_1 being the cash flow at time $t = 1$, c_2 at time $t = 2$, etc.). Again, the present value P of the asset at time $t = 0$ (today) is

$$P_0 = \sum_{t=1}^{\infty} \frac{c_t}{(1 + r)^t} \tag{11}$$

To simplify the model, we start by assuming cash flows are paid continuously (i.e., the time between cash flows approaches zero). In this case, the present value formula becomes

$$P_0 = \int_0^{\infty} c_t e^{-rt} dt \tag{12}$$

as we learned in Section 2.2. In practice, entities both distribute and retain some portion of their after-tax income (profit). We will denote after-tax income as y. We will also denote the portion of after-tax income retained as b and assume that this rate of income retention is fixed over time. The cash flow c_t distributed to investors is then[9]

$$c_t = (1-b)y_t \tag{13}$$

We know from Section 3.1 that profit is expected to grow at rate r (i.e., the cost of capital). If income is fully retained, then we can say the following about income growth:

$$\begin{aligned} y_{t+1} &= y_t(1+r) \\ &= y_t + ry_t \end{aligned}$$

However, if only a portion b of income is retained, then only the retained portion of income should contribute to growth. In this case, we have

$$\begin{aligned} y_{t+1} &= y_t + rby_t \\ &= y_t(1+rb) \end{aligned} \tag{14}$$

In other words, if an entity distributes $(1-b)y_t$ to investors, the price of the entity's equity can't grow at the full rate r. It can only grow by the portion of y_t not distributed, which amounts to the rate of growth br. Notice that Equation 14 is simply a compound interest expression. Define $g = br$ to be the rate at which after-tax income grows continuously. Then back to continuous time, we have

$$y_t = y_0 e^{gt} \tag{15}$$

Substituting Equation 15 into Equation 13 gives us

$$\begin{aligned} c_t &= (1-b)\,y_t \\ &= (1-b)\,y_0 e^{gt} \\ &= c_0 e^{gt} \end{aligned} \tag{16}$$

Substituting Equation 16 back into Equation 12 gives us

$$\begin{aligned} P_0 &= \int_0^\infty c_t e^{-rt}\,dt \\ &= \int_0^\infty c_0 e^{gt} e^{-rt}\,dt \\ &= c_0 \int_0^\infty e^{gt} e^{-rt}\,dt \\ &= c_0 \int_0^\infty e^{-rt+gt}\,dt \\ &= c_0 \int_0^\infty e^{-(r-g)t}\,dt \end{aligned}$$

Integrating, we get

$$\begin{aligned} P_0 &= c_0 \left[\frac{1}{-(r-g)} e^{-(r-g)t} \right]_0^\infty \\ &= c_0 \left[\frac{1}{-(r-g)} e^{-(r-g)\infty} - \frac{1}{-(r-g)} e^{-(r-g)0} \right] \\ &= c_0 \left[0 - \frac{1}{-(r-g)} \right] \\ &= c_0 \left[\frac{1}{r-g} \right] \\ &= \frac{c_0}{r-g} \end{aligned} \tag{17}$$

This gives us the Gordon Growth Model, as originally derived. Solving for the cost of capital r, we see that

$$P_0(r-g)=c_0$$

$$r-g=\frac{c_0}{P_0}$$

$$r=\frac{c_0}{P_0}+g$$

We can say that the cost of capital is equivalent to the dividend yield plus the expected (constant) growth rate. This should be intuitive, as equity returns may take the form of both dividends and capital gains.

Tying this back to capital budgeting discussion from Section 3.1, to equate the marginal project return with the cost of capital, firms should adjust their dividend such that the last project pursued is the project equating these two rates. Following this approach, capital budgeters will theoretically maximize the value of an entity's equity.

Unfortunately, we're not done. While this suffices for assets that make continuous payments, there is a slight change for more realistic payment structures that reflect cash flows in discrete time. Returning to Equation 11, we have

$$P_0=\sum_{t=1}^{\infty}\frac{c_t}{(1+r)^t}$$

Instead of converting to continuous time, we will work through the same procedure from this starting point. Again, we know $g=br$ is the rate at which after-tax income grows, but income will now grow using discrete compounding:

$$y_t=y_0(1+g)^t$$

This contrasts with Equation 15, where we used continuous compounding. We can now state the cash flows to be

$$\begin{aligned} c_t &= (1-b)y_t \\ &= (1-b)y_0(1+g)^t \\ &= c_0(1+g)^t \end{aligned} \tag{18}$$

Plugging Equation 18 into Equation 11, we get

$$\begin{aligned} P_0 &= \sum_{t=1}^{\infty} \frac{c_t}{(1+r)^t} \\ &= \sum_{t=1}^{\infty} \frac{c_0 (1+g)^t}{(1+r)^t} \\ &= c_0 \sum_{t=1}^{\infty} \left(\frac{1+g}{1+r} \right)^t \end{aligned}$$

For ease, let's define $a = \frac{1+g}{1+r}$. Then we have

$$\begin{aligned} P_0 &= c_0 \sum_{t=1}^{\infty} (a)^t \\ &= c_0 \left[a + a^2 + a^3 + a^4 + \cdots \right] \\ &= c_0 a \left[1 + a + a^2 + a^3 + \cdots \right] \end{aligned} \tag{19}$$

To simplify $\left[1 + a + a^2 + a^3 + \cdots\right]$, consider the following exercise:

$$\begin{aligned} x &= 1 + a + a^2 + a^3 + \cdots \\ xa &= 0 + a + a^2 + a^3 + \cdots \\ x - xa &= (1-0) + (a-a) + \left(a^2 - a^2\right) + \left(a^3 - a^3\right) + \cdots \\ x(1-a) &= 1 \\ x &= \frac{1}{1-a} \end{aligned}$$

meaning

$$\left[1 + a + a^2 + a^3 + \cdots\right] = \frac{1}{1-a} \tag{20}$$

Plugging Equation 20 into Equation 19, and remembering $a = \frac{1+g}{1+r}$, we have

$$P_0 = c_0 a\left[1 + a + a^2 + a^3 + \cdots\right]$$

$$= \frac{c_0 a}{1 - a}$$

$$= \frac{c_0 \frac{1+g}{1+r}}{1 - \frac{1+g}{1+r}}$$

Multiplying by $\frac{1+r}{1+r}$, we have

$$P_0 = \left[\frac{c_0 \frac{1+g}{1+r}}{1 - \frac{1+g}{1+r}}\right]\frac{1+r}{1+r}$$

$$= \frac{c_0(1+g)}{(1+r)-(1+g)} \tag{21}$$

$$= \frac{c_0(1+g)}{r-g}$$

This gives us the Gordon Growth Model that is popularly used today. Comparing Equations 11 and 21, we see that this model allows an infinite sum of cash flows growing at a constant rate to be condensed into a single closed-form equation:

$$P_0 = \sum_{t=1}^{\infty} \frac{c_t}{(1+r)^t} = \frac{c_0(1+g)}{r-g}$$

Solving again for r, we get

$$P_0(r-g) = c_0(1+g)$$

$$r - g = \frac{c_0(1+g)}{P_0}$$

$$r = \frac{c_0(1+g)}{P_0} + g$$

Here we have a new expression for the cost of capital when expected cash flows occur at discrete points in the future.

3.2.1 Perpetuity

Following our discussion in the previous section, we know that the present value P_0 of an infinite stream of discretely paid cash flows c growing at rate g is

$$P_0 = \frac{c_0(1+g)}{r-g}$$

Perpetuities are special securities that pay a fixed cash flow at regular intervals forever. Being fixed, the growth rate of these cash flows would be $g=0$. As a result, $c_0 = c_\infty$, meaning the cash flows never change. This allows us to simplify the cash flow notation to c. r remains the opportunity cost of capital. Being a special case of the Gordon Growth Model, a perpetuity may easily be valued as

$$P_0 = \frac{c_0(1+0)}{r-0} = \frac{c}{r} \tag{22}$$

3.2.2 Annuity

Now that we know how to value a never-ending stream of fixed cash flows, we can explore one more simple extension. Annuities are financial products that provide a stream of cash flows (generally annually) for a specified amount of time. They share characteristics with both fixed-rate, annual-pay bonds and perpetuities. They may be viewed as a bond without any return of principle on the last payment, or as a perpetuity that ends.

Finding the value of an annuity is quite easy when it's viewed as a perpetuity that ends. Consider two perpetuities, one starting now P_0 and one starting in five years P_5. Both perpetuities pay the same fixed annual cash flow c and have the same opportunity cost of capital r. From Equation 22, we know that the value of the perpetuity starting now is

$$P_0 = \frac{c}{r} \tag{23}$$

Similarly, we know that the value of the perpetuity starting in five years is

$$P_5 = \frac{c}{r}\frac{1}{(1+r)^5} \tag{24}$$

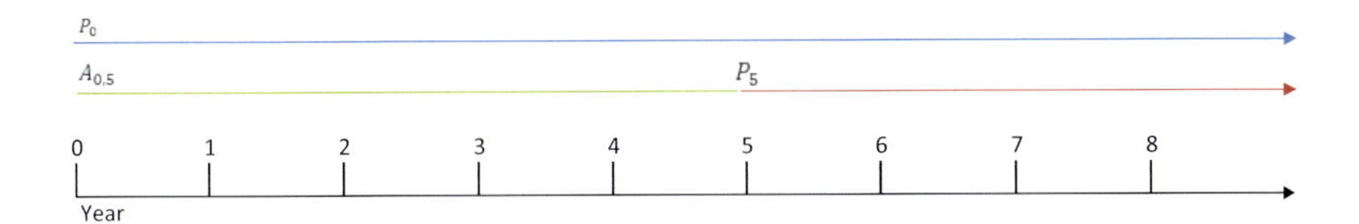

FIGURE 3.1 Present value of an annuity.

The value of this perpetuity is the same as the first, except we are now discounting the value back five years to account for its later starting point.

Say we want to know the value of a five-year annuity $A_{0,5}$ that pays c at the end of each year (starting in one year) and has an opportunity cost of capital r. We can say that

$$A_{0,5} = P_0 - P_5 \tag{25}$$

Why? Well, P_0 is the value of receiving a fixed annual cash flow forever, and P_5 is the value of receiving a fixed annual cash flow forever starting after year five. If we take the difference, we are left only with the value of the fixed cash flows over the first five years. One may visualize this relationship using the illustration in Figure 3.1.

Plugging Equations 23 and 24 into Equation 25, we have

$$\begin{aligned} A_{0,5} &= P_0 - P_5 \\ &= \frac{c}{r} - \frac{c}{r}\frac{1}{(1+r)^5} \\ &= \frac{c}{r}\left[1 - \frac{1}{(1+r)^5}\right] \end{aligned}$$

Generalizing to n years, rather than five, we arrive at the present value of an annuity formula for an annuity paying fixed annual cash flows (with the first payment in one year)[10]:

$$A_{0,n} = \frac{c}{r}\left[1 - \frac{1}{(1+r)^n}\right]$$

3.3 SENSITIVITY

Now that we know how to find the price of a fixed-rate bond, let's explore how the bond price reacts to changes in certain variables. From a risk standpoint, one may examine Equation 6 to understand how the bond price can change. There are three variables that feed into price: c, r, and t. c is fixed in advance for a fixed-rate bond. t, representing when coupons are paid and when principle is returned, is also set out in terms and conditions in advance. These two variables do not provide conduits for unexpected changes in the price B. All unpredictable changes in price must run through the variable r. Recall that r in this context is the YTM.

In this section, we will focus on r. For simplicity, r will be referred to as a "rate" in the remainder of this section and in Sections 3.3.1 and 3.3.2. We will discuss the time variable in Section 4.

Formally writing the bond price B as a function of the rate r, we have $B = f(r)$. Consider a change in rate from r_0 to r_1, resulting in a change in bond price from $f(r_0) = B_0$ to $f(r_1) = B_1$. Let's model B_1 as a Taylor series:

$$\begin{aligned} f(r_1) &= \sum_{d=0}^{\infty} \frac{f^{(d)}(r_0)}{d!}(r_1 - r_0)^d \\ &= f(r_0) + f'(r_0)(\Delta r) + \frac{f''(r_0)}{2}(\Delta r)^2 + \frac{f'''(r_0)}{6}(\Delta r)^3 + \cdots \end{aligned} \tag{26}$$

where superscript (d) refers to the order of the derivative and $r_1 - r_0 = \Delta r$. Note that $f(r_1)$ can be written $B_0 + \Delta B$, where $\Delta B = B_1 - B_0$ is the change in price. Since the red term in Equation 26 is simply equal to B_0, we can simplify Equation 26 to be

$$\begin{aligned} f(r_1) &= f(r_0) + f'(r_0)(\Delta r) + \frac{f''(r_0)}{2}(\Delta r)^2 + \frac{f'''(r_0)}{6}(\Delta r)^3 + \cdots \\ B_0 + \Delta B &= B_0 + f'(r_0)(\Delta r) + \frac{f''(r_0)}{2}(\Delta r)^2 + \frac{f'''(r_0)}{6}(\Delta r)^3 + \cdots \\ \Delta B &= f'(r)(\Delta r) + \frac{f''(r)}{2}(\Delta r)^2 + \frac{f'''(r)}{6}(\Delta r)^3 + \cdots \end{aligned} \tag{27}$$

The goal is to estimate the price change ΔB for a rate change Δr. Equation 27 accomplishes this by using an infinite string of partial derivatives of B with respect to Δr. If every summand were to be included, to infinity, the estimate would be exact.

If a linear approximation were sufficient, then one need only use the blue term in Equation 27, that is:

$$\begin{aligned} \Delta B &\approx f'(r)(\Delta r) \\ &= \frac{dB}{dr}\Delta r \end{aligned} \tag{28}$$

If one were to prefer a bit more precision, then they might include both the blue and pink terms from Equation 27. That is,

$$\begin{aligned} \Delta B &\approx f'(r)(\Delta r) + \frac{f''(r)}{2}(\Delta r)^2 \\ &= \frac{dB}{dr}\Delta r + \frac{1}{2}\frac{d^2B}{dr^2}(\Delta r)^2 \end{aligned} \tag{29}$$

Reliance on Equation 28 implies the use of *duration*, while reliance on Equation 29 implies reliance on duration and *convexity*.

3.3.1 Duration

Let's return to the price of a fixed-rate bond, except for now we're going to assume continuous discounting. Using the tools we've already learned, the price B is elegantly defined in the following form:

$$B = \sum_{i=1}^{n} c_i e^{-rt_i} \tag{30}$$

Here c_i is the ith cash flow (fixed coupon), where the last cash flow c_n includes the return of the bond's principle. Each cash flow occurs at some time t, so t_i represents the time (in years) between today and when cash flow i occurs. The discount rate r represents the YTM of the bond. Discounting is conducted assuming continuous compounding for now, but other compounding frequencies will be explored later in this section.

Since the bond pricing formula is nonlinear, to get the exact price change ΔB for some non-small rate change Δr, one would need to plug both the original and new values of r into Equation 30 and subtract one price from the other. By taking the derivative of the bond price with respect to a change in r, we are getting the change dB for an infinitesimally small change dr. For larger changes Δr, the derivative will provide a linear approximation to ΔB. The approximation, from Equation 28, is written

$$\Delta B \approx \frac{dB}{dr}\Delta r \tag{31}$$

The term $\frac{dB}{dr}$ is the derivative of the bond price B with respect to r. As already mentioned, this represents an amount by which B changes for an infinitesimal change in r. We are then scaling this estimate linearly by the actual amount by which r changes, that is, Δr. This linear scaling provides the linear approximation. Note that, by relying only on duration, one misses detail regarding exposure to changes in the slope or curvature of the yield curve.

CONCEPT REFRESHER 3.2: DERIVATIVES AND DURATION

The derivative of a function represents the change in the function for a change in a variable of the function. The notation is intuitive; think of d as short for *change*. Hence, $\frac{df}{dr}$ literally reads *a change in the function f in response to a change in r*. Formally, a derivative is defined as $\frac{df}{dr} = \lim_{h \to 0} \frac{f(r+h) - f(r)}{h}$. Here h represents the actual change in the underlying variable r. As $h \to 0$, we learn how $f(r)$ responds to an infinitesimally small change in r. Accordingly, the derivative of $f(r)$ can also be considered the *sensitivity* of $f(r)$ to a change in r.

We visualize a derivative in Figure 3.2.

The blue line represents the bond pricing function from Equation 6, which is a nonlinear function of r. The red line represents the approximated value of $f(r)$ using the derivative $f'(r)$ at $r = 0.245$ and scaling by Δr (i.e., $-\Delta r \times f'(0.245)$). It runs tangent to the blue line, intersecting at $r = 0.245$.

Notice that the red line is straight. This is because the first derivative provides only a *linear* approximation to changes in f for changes in r. If we were to zoom in to the point of tangency, at an infinitesimally small level, the red line and blue line are parallel. At this infinitesimally small level, the

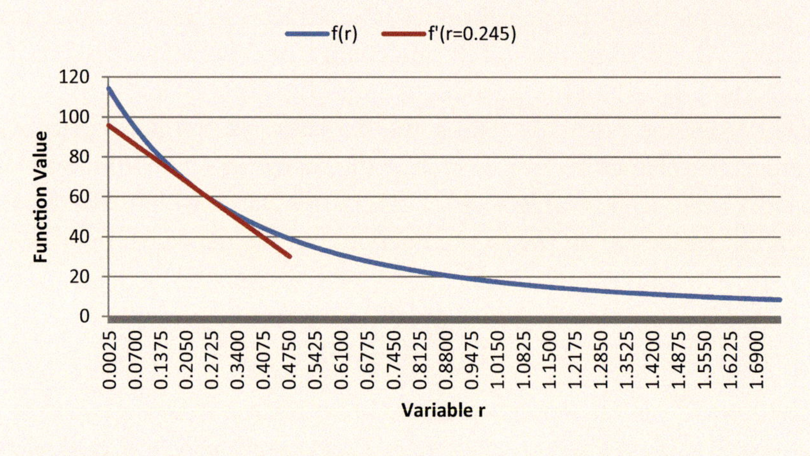

FIGURE 3.2 Derivative visualization.

derivative gives an exact estimate of the change in $f(r)$ for a change in r. As we zoom out, looking at larger changes in r, we see that the red line increasingly separates from the blue line. This shows that the derivative is a decent approximation of the sensitivity of the function for small changes in r, but a lousy one for larger changes in r.

The red line illustrates how one uses duration to approximate the sensitivity of a bond price to changes in r.

Substituting B from Equation 30 into Equation 31, we get

$$
\begin{aligned}
\Delta B &\approx \frac{dB}{dr}\Delta r \\
&= \frac{d}{dr}\sum_{i=1}^{n} c_i e^{-rt_i}\Delta r \\
&= \Delta r \frac{d}{dr}\sum_{i=1}^{n} c_i e^{-rt_i} \\
&= \Delta r\left(\sum_{i=1}^{n} -t_i c_i e^{-rt_i}\right) \\
&= \Delta r\left(-1\sum_{i=1}^{n} t_i c_i e^{-rt_i}\right) \\
&= -\Delta r\sum_{i=1}^{n} t_i c_i e^{-rt_i}
\end{aligned}
\tag{32}
$$

We can now introduce the concept of duration. Duration, broadly speaking, represents the sensitivity of a bond's price B to changes in the discount rate r. *Macaulay duration D* is the earliest definition of duration, which is defined as[11]

$$
D = \frac{\sum_{i=1}^{n} t_i c_i e^{-rt_i}}{B} \tag{33}
$$

Notice that each discounted cash flow $c_i e^{-rt_i}$ in the numerator is being multiplied by the time t_i at which it occurs. Briefly consider this written instead as

$$D = \sum_{i=1}^{n} t_i \frac{c_i e^{-rt_i}}{B}$$

Ultimately, the sum of all $c_i e^{-rt_i}$ terms must equal the price B. So one can think of the term $\frac{c_i e^{-rt_i}}{B}$ as a weight. Specifically, this weight reflects the proportion of the bond price B attributable to each discounted cash flow $c_i e^{-rt_i}$. It follows that D is the weighted average time it will take for a purchaser of the bond to receive the price B back in actual cash flows. It should be clear now why the term 'duration' is used for such a measure.

Multiplying Macaulay duration in Equation 33 by the price of the bond B gives us

$$\begin{aligned} DB &= \frac{B \sum_{i=1}^{n} t_i c_i e^{-rt_i}}{B} \\ &= \sum_{i=1}^{n} t_i c_i e^{-rt_i} \end{aligned} \tag{34}$$

Notice that this is the exact term found in the linear approximation of ΔB from Equation 32. Substituting Equation 34 into Equation 32, we get

$$\begin{aligned} \Delta B &\approx -\Delta r \sum_{i=1}^{n} t_i c_i e^{-rt_i} \\ &= -\Delta r DB \end{aligned}$$

meaning

$$\frac{\Delta B}{B} \approx -\Delta r D$$

This approximation isn't bad, but as mentioned at the beginning of this discussion, the bond price formula from Equation 30 assumes continuous discounting. In reality, bonds pay their coupons at discrete dates (e.g., annually) not continuously. To more accurately estimate the sensitivity of B to changes in r, we must convert from continuous to annual compounding.

We know that an asset earning an annually compounding rate r_a grows by $1+r_a$ over a year. We also know that an asset earning a continuously compounding rate r grows by e^r over a year. Setting these growth expressions equal, we can easily state the annually compounded return in terms of the continually compounded return.[12]

$$1+r_a = e^r$$

Generalizing beyond one year, we raise both sides to the power of t to represent compounding over time.

$$(1+r_a)^t = e^{rt}$$

Taking the natural log of both sides, we get

$$\ln\left[(1+r_a)^t\right] = rt \tag{35}$$

Let's now replace the term rt in the original bond pricing function from Equation 30 with the annual compounding equivalent from Equation 35.

$$\begin{aligned} B &= \sum_{i=1}^{n} c_i e^{-rt_i} \\ &= \sum_{i=1}^{n} c_i e^{-\ln\left[(1+r_a)^{t_i}\right]} \\ &= \sum_{i=1}^{n} \frac{c_i}{e^{\ln\left[(1+r_a)^{t_i}\right]}} \\ &= \sum_{i=1}^{n} \frac{c_i}{(1+r_a)^{t_i}} \end{aligned} \tag{36}$$

Here we have a new form for the price of a bond, where the only difference is that we've switched from continuous compounding to annual compounding.[13] We can now use the derivative of Equation 36 to get a new linear approximation to the change in B for a change in r. Substituting B from Equation 36 into Equation 31, we get

$$
\begin{aligned}
\Delta B &\approx \frac{dB}{dr}\Delta r \\
&= \frac{d}{dr}\sum_{i=1}^{n}\frac{c_i}{(1+r_a)^{t_i}}\Delta r \\
&= \Delta r\frac{d}{dr}\sum_{i=1}^{n}c_i(1+r_a)^{-t_i} \\
&= \Delta r\left(\sum_{i=1}^{n}-t_i c_i(1+r_a)^{-t_i-1}\right) \\
&= \Delta r\sum_{i=1}^{n}\frac{-t_i c_i}{(1+r_a)^{t_i+1}}
\end{aligned}
\tag{37}
$$

Remember that since $(1+r_a)^t$ is just $(1+r_a)(1+r_a)\ldots(1+r_a)$ repeated t times, we can rewrite the denominator from Equation 37 as

$$
(1+r_a)^{t_i+1} = (1+r_a)^{t_i}(1+r_a)
$$

meaning

$$
\begin{aligned}
\Delta B &\approx \Delta r\sum_{i=1}^{n}\frac{-t_i c_i}{(1+r_a)^{t_i+1}} \\
&= \Delta r\sum_{i=1}^{n}(-1)\frac{t_i c_i}{(1+r_a)^{t_i}(1+r_a)} \\
&= -\Delta r\sum_{i=1}^{n}\frac{t_i c_i}{(1+r_a)^{t_i}}\frac{1}{(1+r_a)}
\end{aligned}
$$

Since $\frac{1}{(1+r_a)}$ is just a constant, it can be pulled out

$$
\Delta B \approx -\Delta r\frac{1}{(1+r_a)}\sum_{i=1}^{n}\frac{t_i c_i}{(1+r_a)^{t_i}}
\tag{38}
$$

Consider duration once again from Equation 33, but now replacing the continuously compounded rate with the annually compounded rate from Equation 35, we get

$$\begin{aligned} D &= \frac{\sum_{i=1}^{n} t_i c_i e^{-rt_i}}{B} \\ &= \frac{\sum_{i=1}^{n} t_i c_i e^{-\ln\left[(1+r_a)^{t_i}\right]}}{B} \\ &= \frac{\sum_{i=1}^{n} t_i c_i \left(\frac{1}{e^{\ln\left[(1+r_a)^{t_i}\right]}}\right)}{B} \\ &= \frac{\sum_{i=1}^{n} t_i c_i \left(\frac{1}{(1+r_a)^{t_i}}\right)}{B} \\ &= \frac{\sum_{i=1}^{n} \frac{t_i c_i}{(1+r_a)^{t_i}}}{B} \end{aligned} \tag{39}$$

Once again, by multiplying the duration in Equation 39 by the price of the bond, we get

$$BD = \sum_{i=1}^{n} \frac{t_i c_i}{(1+r_a)^{t_i}} \tag{40}$$

As before, the right-hand side of Equation 40 shows up in the new linear approximation for the change in bond price from Equation 38. Plugging Equation 40 into Equation 38 gives us

$$\begin{aligned} \Delta B &\approx -\Delta r \frac{1}{(1+r_a)} \sum_{i=1}^{n} \frac{t_i c_i}{(1+r_a)^{t_i}} \\ &= -\Delta r \frac{1}{(1+r_a)} BD \\ &= -\frac{\Delta r BD}{1+r_a} \end{aligned}$$

Defining $D_{\text{mod}} = \frac{D}{1+r_a}$ and calling it *Modified duration*, we can say

$$\begin{aligned}\Delta B &\approx -\frac{\Delta r B D}{1+r_a} \\ &= -\Delta r B \frac{D}{1+r_a} \\ &= -\Delta r B D_{\text{mod}}\end{aligned} \tag{41}$$

meaning

$$\frac{\Delta B}{B} \approx -\Delta r D_{\text{mod}}$$

We have now defined Macaulay duration and Modified duration and detailed the link between them. As it turns out, Modified duration can be specified for any compounding frequency using the same procedure. For compounding frequency m (e.g., $m=1$ for annual, $m=2$ for semi-annual, etc.), we have

$$D_{\text{mod}} = \frac{D}{\left(1+\frac{r_m}{m}\right)}$$

Here r_m is the rate that corresponds to the compounding period m. In this way, one may linearly approximate the sensitivity of a bond price B to changes in the rate r for any compounding frequency.

3.3.2 Convexity

To add precision to our estimate of the sensitivity of a bond price B to changes in the rate r, one might wish to use more than just a linear approximation. This can be accomplished with the help of convexity. From Equation 29, we know the estimate for ΔB when including convexity becomes

$$\Delta B \approx \frac{dB}{dr}\Delta r + \frac{1}{2}\frac{d^2 B}{dr^2}(\Delta r)^2$$

Substituting Equation 41 for $\frac{dB}{dr}$, we can now restate this approximation as

$$\Delta B \approx -B D_{\text{mod}} \Delta r + \frac{1}{2}\frac{d^2 B}{dr^2}(\Delta r)^2 \tag{42}$$

We must now focus on the term $\frac{d^2B}{dr^2}$ to include the convexity in our estimate. Returning to our bond price function from Equation 36, we know

$$B = \sum_{i=1}^{n} \frac{c_i}{(1+r_a)^{t_i}}$$

where B is the bond price, c_i is the cash flow occurring at time t_i, and r_a is the annually compounded rate. We also know from Equation 37 that

$$\frac{dB}{dr} = \sum_{i=1}^{n} \frac{-t_i c_i}{(1+r_a)^{t_i+1}} \tag{43}$$

To calculate the convexity, we must once again find the derivative of Equation 43 with respect to a change in r. We have

$$\begin{aligned}
\frac{d^2B}{dr^2} &= \frac{d}{dr}\sum_{i=1}^{n} \frac{-t_i c_i}{(1+r_a)^{t_i+1}} \\
&= \frac{d}{dr}\sum_{i=1}^{n} -t_i c_i (1+r_a)^{-(t_i+1)} \\
&= \sum_{i=1}^{n} -(t_i+1)(-t_i c_i)(1+r_a)^{-(t_i+1)-1} \\
&= \sum_{i=1}^{n} \frac{(t_i+1)(t_i c_i)}{(1+r_a)^{t_i+2}} \\
&= \frac{1}{(1+r_a)^2}\sum_{i=1}^{n} \frac{(t_i^2+t_i)c_i}{(1+r_a)^{t_i}}
\end{aligned}$$

Defining the *modified convexity* C_{mod} to be

$$C_{\text{mod}} = \frac{1}{B(1+r_a)^2}\sum_{i=1}^{n} \frac{(t_i^2+t_i)c_i}{(1+r_a)^{t_i}}$$

we can then restate Equation 42 as

$$\Delta B \approx -BD_{\text{mod}}\Delta r + \frac{1}{2}\frac{d^2B}{dr^2}(\Delta r)^2$$

$$= -BD_{\text{mod}}\Delta r + \frac{1}{2}BC_{\text{mod}}(\Delta r)^2$$

meaning

$$\frac{\Delta B}{B} \approx -D_{\text{mod}}\Delta r + \frac{1}{2}C_{\text{mod}}(\Delta r)^2$$

NOTES

1 *Zero rates* are rates on zero-coupon bonds. Zero-coupon bonds are bonds that pay a coupon rate of 0% but are sold at a discount from par. If a zero-coupon bond is sold at 95 with a par value of 100 and one year remaining to maturity, the return on the bond is $5/95 = 5.3\%$. This serves as the one-year zero rate.
2 Much of this section is adapted from Gordon and Shapiro (1956).
3 We use the *marginal* project return because the marginal project represents a feasible use of available funds. Inframarginal projects (i.e., projects already in action) might have higher returns, which is why they were chosen first, but they're not useful in assessing an opportunity cost since a new investment could not attain such returns.
4 The concept of the IRR is explained in more detail in Concept Refresher 3.1.
5 Both the marginal project return and the expected rate of profit are categorically different than the risk-free rate, in that they represent returns that are not guaranteed (i.e., risky returns). As we will see in Section 9, viewing risky returns as the sum of the risk-free return and a risky spread can yield useful information.
6 As an aside, this is the motivation behind why the yield to maturity of a bond includes reinvestment return in addition to capital gains and coupon payments. See Concept Refresher 3.1 for more on the opportunity cost of capital as the yield to maturity.
7 This is also referred to as having positive NPV.
8 Much of this section is adapted from Gordon and Shapiro (1956).
9 Here c_t may also be referred to as the *dividend* at time t.
10 One may hear about a slightly different structure called an *annuity due*, which just means the payments are made at the beginning of each year rather than the end. If the first payment is carved out and separated from the remaining payments, an annuity due is identical to a standard annuity plus a payment at the beginning of the period. The present value of an annuity due $A_{0,n}^{\text{due}}$ is then simply $A_{0,n} + c$, where c is still the annual cash flow.

11 Macaulay (1938). Macaulay duration is named after Frederick Macaulay. An academic financier, his research interests included income distribution, interest rate modeling, and security pricing. His contribution of Macaulay duration may be found in his book *Some Theoretical Problems Suggested by the Movements of Interest Rates, Bond Yields and Stock Prices in the United States since 1856* written in 1938.

12 An interest rate with any assumed compounding frequency can easily be converted into an interest rate with a different compounding frequency by simply setting the future values equal. For example, if we know the annually compounded rate r_a and want to convert from annual compounding to semi-annual compounding to calculate the equivalent semi-annually compounded rate r_s, one simply sets $(1+r_a)^t = \left(1+\frac{r_s}{2}\right)^{2t}$ and then solves for r_s.

13 Notice that Equation 36 matches Equation 6 from Section 3, except now we include the subscript *a* to specify that r_a is an annually compounded rate.

SECTION 4

Time Series Processes[1]

NOW THAT WE HAVE begun exploring the sensitivity of modeled security prices to changes in variables, let's talk about sensitivity to a particular variable: *time*. Since time is never constant, every security's price changes with time. As time passes, one observes a sequence of successive security values, forming a sort of "price path."

To understand the price path of a financial security, we often talk of the process generating that path. Figure 4.1 shows a price chart for Apple, Inc. (ticker: AAPL) over a randomly chosen time period.

If you were to list out each day's price, you would have a series of numbers. This is known as a time series. Notice, in this particular case, the time series is increasing over time. This is actually a problem from a statistical standpoint. If you take the average price over time, it will continually go up! When the statistics (mean, variance, etc.) of a time series change over time, the time series is said to be *non-stationary*. Statisticians prefer stationarity for a host of reasons, mostly because stationary data satisfies key assumptions for statistical tests. To turn a non-stationary time series into a stationary one, it is often sufficient to difference the data. This means that instead of working with a financial security's price, we work with the *change* in its price. Differencing Apple's daily prices from Figure 4.1, we get Figure 4.2.

Notice how the change in price fluctuates over a somewhat stable level. We can refer to Apple's stock price at time t as y_t, and we can refer to the change in Apple's stock price as $\Delta y_t = y_t - y_{t-1}$. So y_0 would be the price at the very beginning of the period, and we commonly refer to y_T as the price at the very end of the period. y_t will then represent any price in the time

DOI: 10.1201/9781032687650-4

FIGURE 4.1 AAPL price chart.

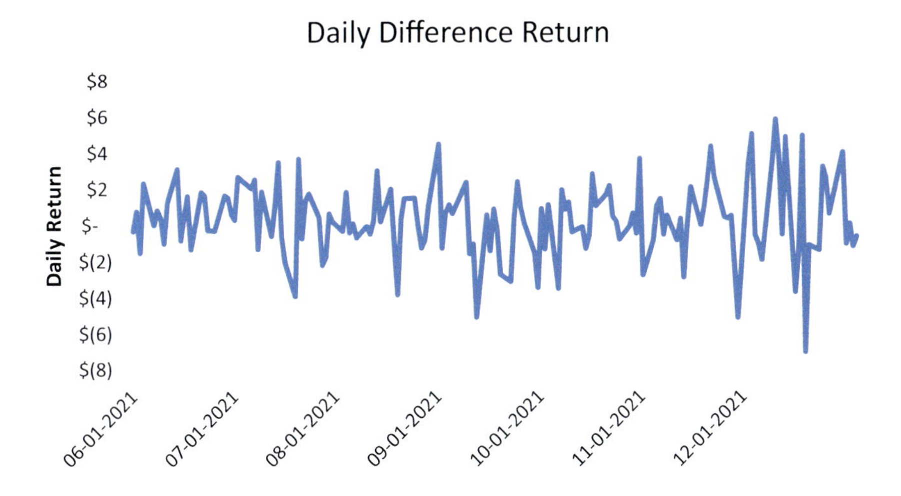

FIGURE 4.2 AAPL difference return.

period between $t = 0$ to $t = T$. Similarly, Δy_t represents the change in price for any intervening period between $t = 0$ and $t = T$.

There are two broad ways to categorize time series processes: deterministic and stochastic. The path of a deterministic time series is known in advance, with no random fluctuations intervening. Stochastic processes may reflect partially deterministic time series but include some form of random noise that changes the trajectory of the path. We'll review each in the following sections.

4.1 DETERMINISTIC PROCESSES

Let's start with a simple financial security S that earns a deterministic, continuous risk-free rate r. The security is held for some time t. If S_0 is the initial price, then the price of the security at time t is

$$S_t = S_0 e^{rt} \tag{44}$$

The return per unit of time is calculated by taking the partial derivative with respect to t:

$$\begin{aligned}\frac{dS_t}{dt} &= \frac{d}{dt} S_0 e^{rt}\\ &= S_0 \frac{d}{dt} e^{rt}\\ &= S_0 \left[re^{rt} \right]\\ &= rS_0 e^{rt}\\ &= rS_t\end{aligned}$$

Note that the return per unit of time is proportional to the price of the security (i.e., $\frac{dS_t}{dt} \propto S_t$). Distributing the time increment dt, we get

$$\frac{dS_t}{dt} = rS_t$$

$$dS_t = S_t r dt$$

This exercise suggests that for a given increment of time, the price of the security changes by rS_t (scaled to the size of the time increment dt). Since both r and S_t are known at time t, security S is said to follow a deterministic process.

4.2 STOCHASTIC PROCESSES

Rather than just receiving a deterministic return r over time, security S now receives an additional stochastic return ε over the same period. The future price is now modeled as

$$S_t = S_0 e^{(r+\varepsilon_t)t}$$

where ε_t is best viewed as a noise variable. To properly represent random noise, ε_t is defined as a stochastic process with independent and stationary increments.[2] Stationary increments means the increment $\varepsilon_{t+i} - \varepsilon_t$ follows the same probability distribution, with constant parameters, for all times t where i is also an arbitrary amount of time. In other words, the distribution of the noise variable is unchanged across time. Independent increments means the increments $\varepsilon_i - \varepsilon_0$ and $\varepsilon_{t+j} - \varepsilon_t$, $0 \leq i \leq t$ are independent for all arbitrary times i, j, t. In addition, to be true noise, ε_t should have a mean of zero (i.e., $E(\varepsilon_t) = 0$ for all t).

Once again calculating the return per unit of time, we have

$$\begin{aligned}
\frac{dS_t}{dt} &= \frac{d}{dt} S_0 e^{(r+\varepsilon_t)t} \\
&= S_0 \frac{d}{dt} e^{(r+\varepsilon_t)t} \\
&= S_0 \left[(r+\varepsilon_t) e^{(r+\varepsilon_t)t} \right] \\
&= (r+\varepsilon_t) S_0 e^{(r+\varepsilon_t)t} \\
&= (r+\varepsilon_t) S_t \\
&= rS_t + \varepsilon_t S_t
\end{aligned} \tag{45}$$

In the current form, there is no process for ε_t that satisfies the independent and stationary increments requirements. ε_t, being a random variable, does not have a continuous path and under certain conditions cannot even be a measurable function. To overcome this difficulty, we must leave the world of infinitesimals and approximate this same model in discrete time. In other words, we shift from dt to Δt. $\Delta S_t = S_t - S_{t-i}$ now represents the change in price of the security from discrete time $t-i$ to discrete time t. Let's rewrite the return per unit of time from Equation 45 in discrete time:

$$\frac{\Delta S_t}{\Delta t} = rS_t + \varepsilon_t S_t$$

As with the deterministic case, the return per unit of time is still proportional to the price of the security. Distributing the discrete time increment, we have

$$\Delta S_t = rS_t \Delta t + \varepsilon_t S_t \Delta t \tag{46}$$

We may now take ε_t and Δt together as $\varepsilon_t \Delta t = \Delta W_t$. While ε_t cannot meet the independent and stationary increments requirements, the process W_t can. As we will discuss in the following section, a Wiener process is a perfect candidate for W_t, as it meets all independent and stationary increments requirements and has a continuous path. We can now rewrite Equation 46 as

$$\Delta S_t = rS_t\Delta t + S_t\Delta W_t$$

In contrast to the deterministic case, the return per unit of time now includes unknown variables. While r and S_t are known at time t, ΔW_t is not. Since the movement implied by ΔW_t is random, security S is now said to follow a stochastic process.

Before continuing, note the general structure of this stochastic process. There are two distinct terms. As already discussed, the first term $rS_t\Delta t$ is deterministic. This term is commonly referred to as the *drift*. The second term $S_t\Delta W_t$ is referred to as *noise*. The combination of these two terms provides the flexibility to model all sorts of time series processes. As we'll see in the coming sections, different variants of this process can be utilized when pricing certain financial assets.

CONCEPT REFRESHER 4.1: PROBABILITY VS. CUMULATIVE PROBABILITY

When we say *distribution*, we are casually referring to the probability distribution underlying some process (i.e., the probability that a random variable will take certain values). Such a concept may be thought of in two ways: through the probability mass function (PMF) or the cumulative distribution function (CDF).

The CDF is a function describing the probability that a variable is less than or equal to some threshold, hence the word *cumulative* in the name. Using Φ to denote the CDF, one might say $\Phi(x) = P(X \leq x)$ where X is the random variable and x is any actual value that X can take. Say our domain is $x \in \{1,\ldots,10\}$, and the probability of X taking any of these values is equal. In other words, there is a 10% probability of drawing $x = 1$, a 10% probability of drawing $x = 2$, and so on. Then $\Phi(5) = P(X \leq 5) = 50\%$. Figure 4.3 provides a visualization.

The PMF, denoted as ϕ, is what is being summed to attain the CDF. In the previous example given, the probability of drawing $x = 1$ was given as 10% (i.e., $\phi(1) = P(X = 1) = 10\%$). Since all variables in this example have equal probability, the PMF would trivially look as in Figure 4.4:

If instead, we were interested in the probability that X takes the values 3, 4, or 5, we can write this as $P(2 < X \leq 5) = \Phi(5) - \Phi(2) = \sum_{x=3}^{5}\phi(x)$, demonstrating that probabilistic statements may be expressed in terms of either the PMF or the CDF.

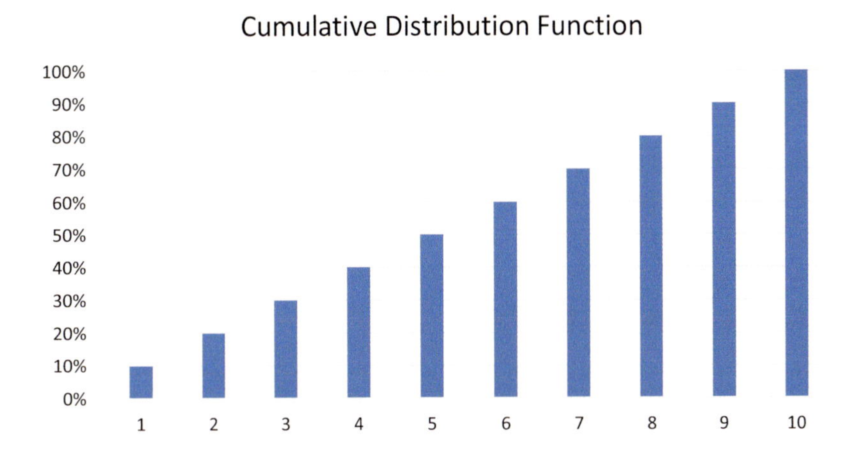

FIGURE 4.3 Cumulative distribution function.

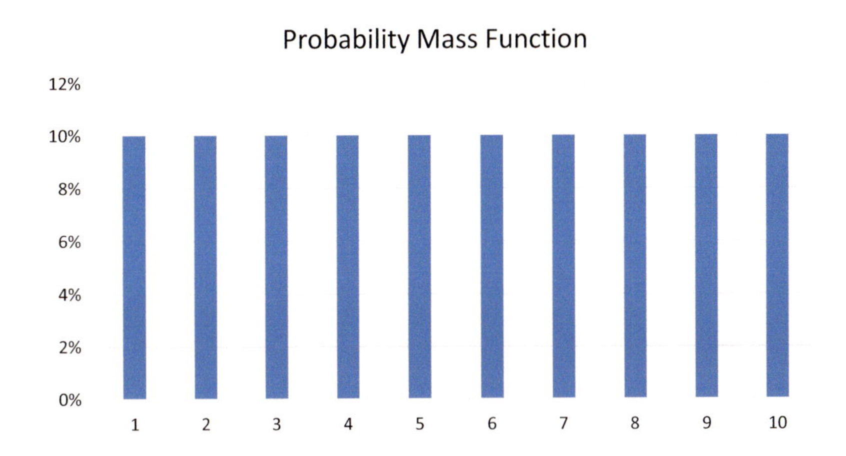

FIGURE 4.4 Probability mass function.

4.2.1 Random Walk

To get a better feel for the stochastic term in stochastic processes, specifically the Wiener process, let's first consider a process that is entirely random (i.e., zero drift). One example of this is a simple random walk. This may be written as

$$y_t = y_{t-1} + \varepsilon_t, \text{ or}$$
$$\Delta y_t = \varepsilon_t \tag{47}$$

where y_t may be thought of as a price at time t and the random variable ε_t is similar to the noise term from Section 4.2, except it now has the piecewise form $\varepsilon_t = \begin{cases} +1 & \text{with 50\% probability} \\ -1 & \text{with 50\% probability} \end{cases}$. In this model, the security price goes up or down by one unit at each point in time. For now, we'll assume the price can only take discrete integer values, and that time moves in discrete steps of one unit (i.e., $\Delta t = 1$). Using a starting price of $y_0 = 0$ and generating five sets of 100 iterations for ε_t, we can plot five different series using this random walk model (Figure 4.5):

At any particular time t, the price $y_t = y_0 + \sum_{i=1}^{t} \varepsilon_i$. Let's see what we can learn about the process in a single period. Taking the expectation of Equation 47 gives us

$$E(\Delta y_t) = E(\varepsilon_t) = 0.5 \times 1 + 0.5 \times (-1) = 0$$

The variance of Equation 47 is calculated as

$$\text{Var}(\Delta y_t) = \text{Var}(\varepsilon_t) = E(\varepsilon_t^2) - E(\varepsilon_t)^2 = E(\varepsilon_t^2) - 0 = 1$$

since $E(\varepsilon_t^2) = 0.5 \times 1^2 + 0.5 \times (-1)^2 = 1$. We can say that ε_t has a piecewise distribution with a mean of zero and a variance of one.

As it turns out, with some rescaling, the random walk converges to a Wiener process as the time increment Δt approaches zero.[3] To get an intuition for this convergence, let's reconsider the random walk through a

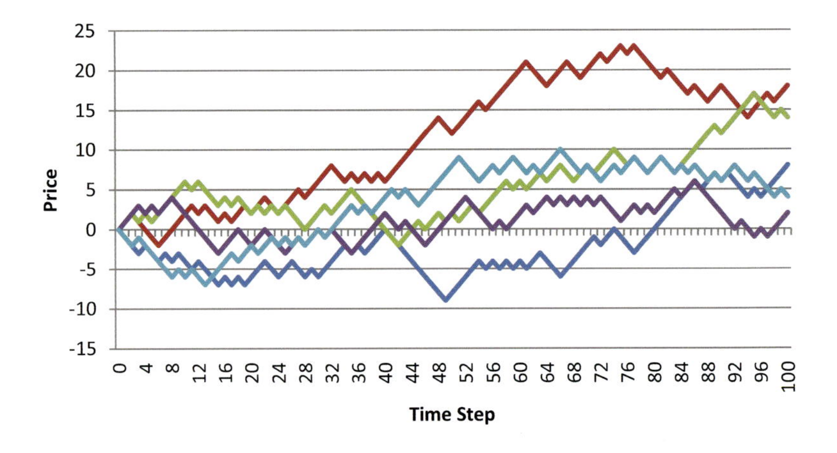

FIGURE 4.5 Random walk iterations.

more general lens. As before, y_t denotes a price at time t. Instead of increasing or decreasing by exactly 1, y now moves in steps of size δ either up or down every unit of time Δt. This means there are $t/\Delta t$ units of time between time 0 and time t. To determine whether y moves up or down each time increment, we define the random variable

$$\varepsilon_i = \begin{cases} 1, & \text{if up move} \\ -1, & \text{if down move} \end{cases}$$

where each ε_i is independent. And once again, $P(\varepsilon_i = 1) = P(\varepsilon_i = -1) = 0.5$. Formally, we have

$$y_t = \delta\varepsilon_1 + \cdots + \delta\varepsilon_{t/\Delta t} = \delta(\varepsilon_1 + \cdots + \varepsilon_{t/\Delta t}) \tag{48}$$

We know already that $E(\varepsilon_i) = 0$ and $\text{Var}(\varepsilon_i) = 1$ from our analysis of Equation 47. Finding the expectation and variance of y_t in Equation 48, we have

$$\begin{aligned} E(y_t) &= E[\delta(\varepsilon_1 + \cdots + \varepsilon_{t/\Delta t})] \\ &= \delta E(\varepsilon_1 + \cdots + \varepsilon_{t/\Delta t}) \\ &= \delta\left[\sum_{i=1}^{t/\Delta t} E(\varepsilon_i)\right] = 0 \end{aligned} \tag{49}$$

$$\begin{aligned} \text{Var}(y_t) &= \text{Var}(\delta(\varepsilon_1 + \cdots + \varepsilon_{t/\Delta t})) \\ &= \delta^2 \text{Var}(\varepsilon_1 + \cdots + \varepsilon_{t/\Delta t}) \\ &= \delta^2\left[\sum_{i=1}^{t/\Delta t} \text{Var}(\varepsilon_i)\right] \text{ by independence} \\ &= \delta^2 \frac{t}{\Delta t} \end{aligned} \tag{50}$$

As mentioned previously, for the random walk to converge to a Wiener process, we must switch to treating time and price as continuous rather than discrete. Instead of using the discrete time steps $\Delta t = 1$, we take the limit as $\Delta t \to 0$. These new infinitely small increments (denoted dt) are known as *infinitesimals*.

Similarly, we must let the step size δ approach zero. However, we must be careful how we structure this limit. If we simply set $\delta = \Delta t$, for example, then both $E(y_t)$ and $\mathrm{Var}(y_t)$ converge to zero as $\Delta t \to 0$. Instead, we will strategically assign the step size to be

$$\delta = \sigma\sqrt{\Delta t} \tag{51}$$

where σ is an arbitrary positive constant. Using the step size in Equation 51, we may now restate the expectation and variance of y_t from Equations 49 and 50, respectively.

$$E(y_t) = 0 \tag{52}$$

$$\begin{aligned} \mathrm{Var}(y_t) &= \delta^2 \frac{t}{\Delta t} \\ &= \left(\sigma\sqrt{\Delta t}\right)^2 \frac{t}{\Delta t} \\ &= \sigma^2 \Delta t \frac{t}{\Delta t} = \sigma^2 t \end{aligned} \tag{53}$$

In the special case where $\sigma = 1$, the variance becomes $\mathrm{Var}(y_t) = t$. Switching now to the change in value each time increment, we can express Equation 48 recursively:

$$\begin{aligned} y_t &= \delta(\varepsilon_1 + \cdots + \varepsilon_{t/\Delta t}) = \delta\left(\varepsilon_1 + \cdots + \varepsilon_{(t-\Delta t)/\Delta t}\right) + \delta\varepsilon_{t/\Delta t} \\ &= y_{t-\Delta t} + \delta\varepsilon_{t/\Delta t} \end{aligned} \tag{54}$$

Using Equations 51 and 54 and assuming $\sigma = 1$, we have

$$\begin{aligned} \Delta y_t &= y_t - y_{t-\Delta t} \\ &= y_{t-\Delta t} + \delta\varepsilon_{t/\Delta t} - y_{t-\Delta t} \\ &= \delta\varepsilon_{t/\Delta t} \\ &= \varepsilon_{t/\Delta t}\sqrt{\Delta t} \end{aligned} \tag{55}$$

This is similar to the expression for the random walk from Equation 47, except now we have more flexibility with the step size δ and the time increment Δt. Finding expectation and variance again, we have

$$E(\Delta y_t) = E\left(\varepsilon_{t/\Delta t}\sqrt{\Delta t}\right) = 0 \tag{56}$$

$$\mathrm{Var}(\Delta y_t) = \mathrm{Var}\left(\varepsilon_{t/\Delta t}\sqrt{\Delta t}\right) = \Delta t\,\mathrm{Var}(\varepsilon_{t/\Delta t}) = \Delta t \tag{57}$$

The results from Equations 52, 53, 56, and 57 show up again in Section 4.2.2, as they are the same expected value and variance of a price following a Wiener process.

CONCEPT REFRESHER 4.2: DISCRETE VS. CONTINUOUS

Discrete distributions have a set of distinct possible values. Consider a *discrete* distribution consisting of all integers between –5 and 5. Figure 4.6 shows how a fictitious PMF might look.

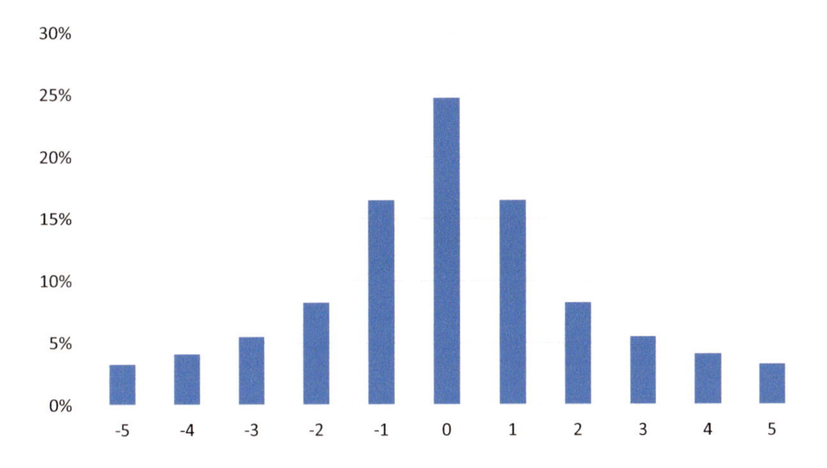

FIGURE 4.6 Discrete distribution.

Each integer, including zero, has a probability associated with it, with the probability decreasing as the distance from zero increases. The probability of any non-integer is zero.

A *continuous* distribution does not have a distinct set of possible values. Rather, a continuous distribution can include integers and all numbers in between, out to an infinite number of decimals. Consider the same example but as a continuous distribution. The probability *density* function (PDF) would now look as follows (Figure 4.7).

Notice the distribution covers the same domain but now includes all possible numbers between integers. Since there is an infinite number of real numbers between any two values on a continuous scale, the probability of any one single number is 0% (since $\int_a^a f(x)dx = 0$). Thus, with continuous distributions, probability must be measured over ranges. For example, the probability between –1 and 1 is 68.3% in this density function (which is just the standard normal density), even though the probability of any individual number between –1 and 1 is 0%.

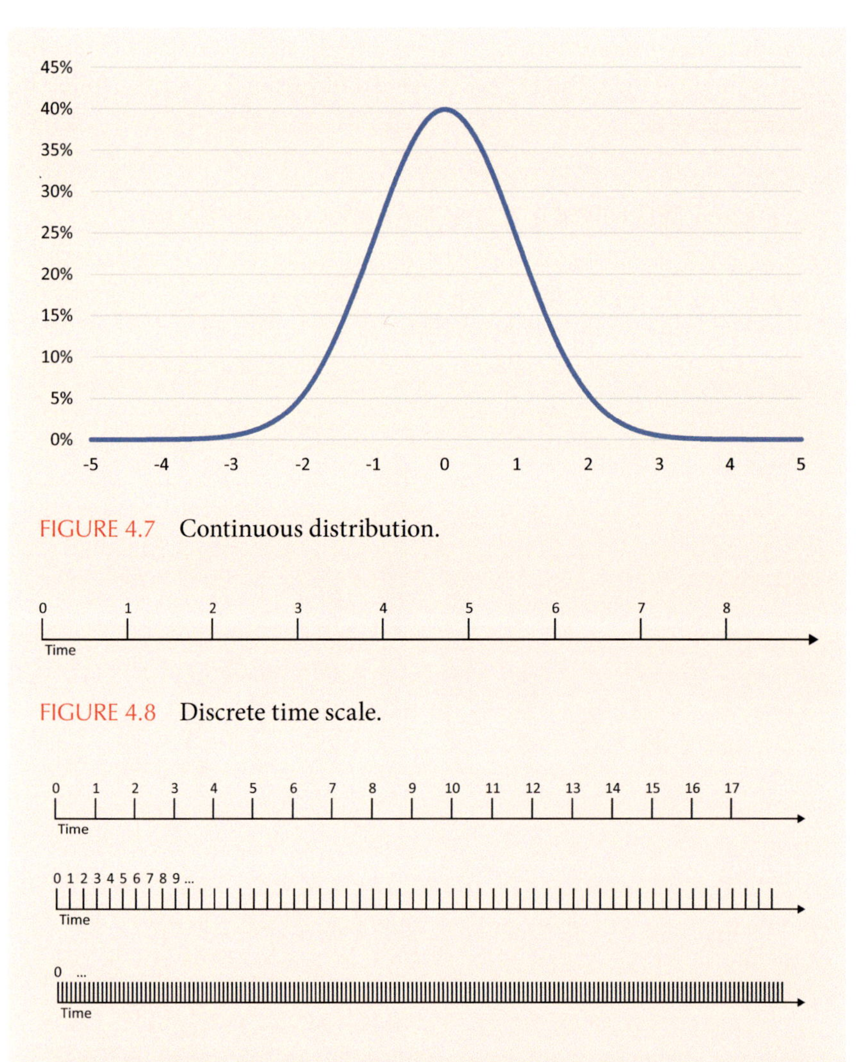

FIGURE 4.7 Continuous distribution.

FIGURE 4.8 Discrete time scale.

FIGURE 4.9 Discrete to continuous time scale.

The same concept applies to a time series. In discrete time, each time series value corresponds to a particular point in time. These points are generally separated by a fixed interval. The timeline in Figure 4.8 emphasizes discrete points separated by an interval of unit one.

To recover a continuous time scale, start reducing the interval between discrete points (Figure 4.9).

As the interval approaches zero, the timeline becomes continuous.

Note that for continuous distributions, the CDF is no longer the sum of the PDF. Rather, the CDF is now the *integral* of the PDF, making the PDF the derivative of the CDF.

4.2.2 Wiener Process

Before introducing the Wiener process, we assign a continuous probability distribution to ε rather than the discrete piecewise functions from Equations 47 and 48. A common probability distribution to use here is the normal distribution. The normal distribution is the bell-shaped distribution one is likely to remember from high school statistics. Figure 4.10 provides an illustration of the normal distribution.

The normal distribution is fully defined by two parameters, the mean and variance. When you hear that a variable has a standard normal distribution, this suggests these two parameters have been standardized to have a mean of zero and a standard deviation of one. If ε_t follows a standard normal distribution, we write $\varepsilon_t \sim N(0,1)$.[4]

With the new definition of ε_t, we can describe the extension of the random walk known as a standard Brownian motion (after Robert Brown[5]). It is also known as a Wiener process (after Norbert Wiener[6]).

Formally, we have[7]

$$\Delta W_t = \varepsilon_t \sqrt{\Delta t}$$

A Wiener process W_t states that a small change $\Delta W_t = W_t - W_{t-\Delta t}$ satisfies certain properties:

1. $\Delta W_t = \varepsilon_t \sqrt{\Delta t}$, where ε_t is a standard normal random variable $\left(\text{i.e., } \varepsilon_t \sim N(0,1)\right)$

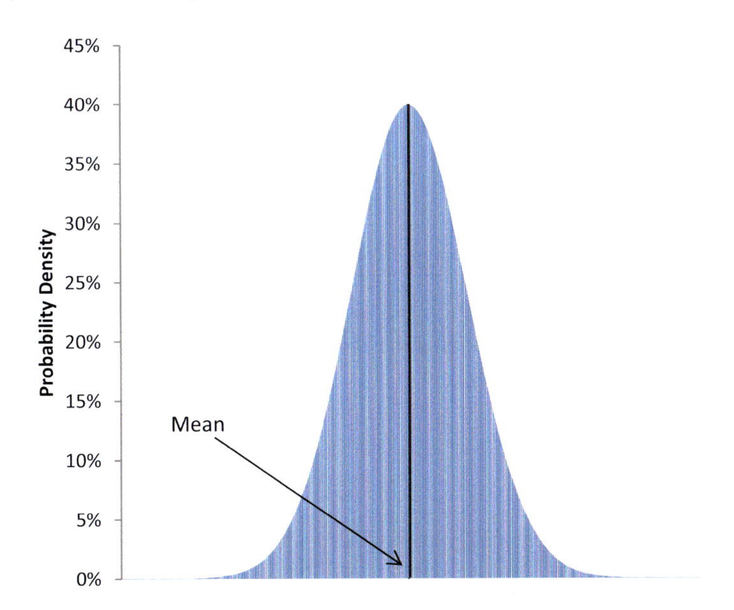

FIGURE 4.10 Normal distribution.

2. Future increments ΔW_u for all $u > t$ are independent of past increments ΔW_s for all $s \leq t$

The first property is known as Gaussian increments. Gaussian is another word for normal, which stems from the fact that the normal distribution was derived by Carl Friedrich Gauss.[8] Taking expectation of the Wiener process, we can see that

$$E(\Delta W_t) = E\left(\varepsilon_t \sqrt{\Delta t}\right) = E(\varepsilon_t)\sqrt{\Delta t} = 0$$

since $E(\varepsilon_t) = 0$. The variance is

$$\begin{aligned}
\text{Var}(\Delta W_t) &= E\left(\Delta W_t^2\right) - E(\Delta W_t)^2 \\
&= E\left(\varepsilon_t^2 \sqrt{\Delta t}^2\right) - E\left(\varepsilon_t \sqrt{\Delta t}\right)^2 \\
&= E\left(\varepsilon_t^2\right)\Delta t - 0 = \Delta t
\end{aligned}$$

since $E\left(\varepsilon_t^2\right) = 1$.

The second property is known as the Markov property, in which the next value W_{t+1} depends only on the present state W_t and not on the sequence that precedes it $\left\{W_j\right\}$ for $j < t$.

Assume that the process starts with an initial value of $w_0 = 0$. Then $W_t - w_0$ represents the sum of many tiny increments. If $n = t / \Delta t$ is the total number of time increments, then

$$W_t - w_0 = W_{n\Delta t} - w_0 = \sum_{i=1}^{n} \Delta W_i = \sum_{i=1}^{n} \varepsilon_i \sqrt{\Delta t} = \sqrt{\Delta t} \sum_{i=1}^{n} \varepsilon_i$$

Because of the Gaussian increment and Markov properties, we have

$$E(W_t - w_0) = \sqrt{\Delta t} \sum_{i=1}^{n} E(\varepsilon_i) = 0$$

and variance

$$\begin{aligned}
\text{Var}(W_t - w_0) &= \text{Var}\left(\sqrt{\Delta t} \sum_{i=1}^{n} \varepsilon_i\right) \\
&= \sqrt{\Delta t^2} \sum_{i=1}^{n} \text{Var}\left(\varepsilon_i^2\right) \\
&= \Delta t \sum_{1}^{n} 1 = n\Delta t = t
\end{aligned}$$

since, again, $E\left(\varepsilon_i^2\right)=1$, $E(\varepsilon_i)=0$, and $\sum_1^n 1=n$. So $(W_t-w_0)\sim N(0,t)$, meaning a Wiener process has zero drift and a variance proportional to the time interval Δt.

This, unfortunately, is not general enough. Financial securities, especially stocks, grow in more complex ways. Fortunately, it's easy to generalize the Wiener process into a process that has expected drift μ (meaning the mean change is proportional to μ) and variance rate σ^2 (meaning the variance is proportional to σ^2). μ and σ^2 can be thought of as scaling the drift and variance terms, respectively. Without them, the Wiener process is limited to the properties of a standard normal variable. Note that the square root of the variance rate, σ, is often referred to as the *volatility rate*.

Using common notation, the *generalized Wiener process* is written

$$dy_t=\mu\ dt+\sigma\ dW_t \tag{58}$$

where $dW_t=\varepsilon_t\sqrt{dt}=\lim_{\Delta t\to 0}\varepsilon_t\sqrt{\Delta t}$ is a Wiener process. A discrete version of this generalized model is

$$\Delta y_t=\mu\Delta t+\sigma\varepsilon_t\sqrt{\Delta t}$$

If we are interested in the period from time 0 to time t, then $\Delta t=t-0=t$ and we write

$$y_t-y_0=\mu t+\sigma\varepsilon_t\sqrt{t}$$

Taking expectation gives

$$E\left(y_t-y_0\right)=E\left(\mu t+\sigma\varepsilon_t\sqrt{t}\right)=\mu t+\sigma E(\varepsilon_t)\sqrt{t}=\mu t$$

since $E(\varepsilon_t)=0$. The variance is

$$\begin{aligned}\mathrm{Var}\left(y_t-y_0\right)&=\mathrm{Var}\left(\mu t+\sigma\varepsilon_t\sqrt{t}\right)\\&=E\left(\mu^2t^2+\sigma^2\varepsilon_t^2\sqrt{t}^2\right)-E\left(\mu t+\sigma\varepsilon_t\sqrt{t}\right)^2\\&=\mu^2t^2+\sigma^2E\left(\varepsilon_t^2\right)\sqrt{t}^2-\mu^2t^2=\sigma^2t\end{aligned} \tag{59}$$

Notice that both the mean and variance scale linearly with time t. Taking the square root of the variance gives the volatility $\sigma\sqrt{t}$, which scales with the square root of time. When we say *scales* here, we're referring to the way in which the mean and volatility change as t (time) increases.

4.2.3 Ito Process

Now that we understand the form and properties of a stochastic process, we must introduce a special type of stochastic process that will be useful in a later section. This is known as an Ito process, and it generalizes the generalized Wiener process even further, allowing the drift and volatility to be time-dependent. An Ito process is formally written as[9]

$$dy_t = \mu(y_t, t)dt + \sigma(y_t, t)dW_t \tag{60}$$

Ito developed one of the most important theorems in mathematical finance, now known as Ito's lemma.

Consider two variables, y and $G(y)$. Both y and G represent prices, but for categorically different objects. The object represented by y has no contingencies. Its price is observable without any other input. It might be oil, gold, a stock, or even an interest rate. G, on the other hand, has contingencies. It represents a *derivative* of y, in that G is a function of y.[10] As y changes so does G, following some functional relationship.

Suppose now we wanted to specify a process detailing the evolution of both y and G over time. Perhaps y follows a deterministic or stochastic process of a form similar to that already discussed. If so, how do we model G? Ito's lemma provides an answer. Specifically, Ito's lemma allows us to specify a process for G given that we know the process for y. The lemma can be derived as follows.

Let the price $G(y)$ be a differentiable function of the price y. The Taylor expansion of $G(y)$ for a change in y is written

$$\Delta G \equiv G(y + \Delta y) - G(y) = \frac{\partial G}{\partial y}\Delta y + \frac{1}{2}\frac{\partial^2 G}{\partial y^2}(\Delta y)^2 + \frac{1}{6}\frac{\partial^3 G}{\partial y^3}(\Delta y)^3 + \cdots$$

Taking the limit as $\Delta y \to 0$ (so switching from Δ to d) and ignoring higher order terms of Δy (i.e., ignoring terms with an exponent greater than one), we get $dG = \frac{\partial G}{\partial y}dy$.

Now consider a discrete version of the Ito process:

$$\Delta y = \mu \Delta t + \sigma \varepsilon_t \sqrt{\Delta t} \tag{61}$$

Again, $\varepsilon \sim N(0,1)$. Refer back to the Ito process from Equation 60. Both the drift term and the variance term are functions of two variables, t (time) and y_t (stock price). Restating ΔG as a function of both variables, we have

$$\Delta G = \frac{\partial G}{\partial t}\Delta t + \frac{\partial G}{\partial y}\Delta y + \frac{1}{2}\frac{\partial^2 G}{\partial y^2}(\Delta y)^2 + \frac{1}{2}\frac{\partial^2 G}{\partial t^2}(\Delta t)^2 + \frac{1}{2}\frac{\partial^2 G}{\partial t \partial y}\Delta t \Delta y + \cdots \tag{62}$$

Taking the limit as $\Delta t \to 0$ and $\Delta y \to 0$ and assuming that higher order terms of Δt may be ignored[11] gives us $dG = \frac{\partial G}{\partial t} dt + \frac{\partial G}{\partial y} dy$.

However, this is not quite accurate. The terms in red in Equation 62 really do contain higher order terms only. Consider $\Delta t \Delta y$, the last term in red. Plugging Equation 61 for Δy, we have

$$\Delta t \Delta y = \Delta t\left(\mu \Delta t + \sigma \varepsilon_t \sqrt{\Delta t}\right) = \mu \Delta t^2 + \sigma \varepsilon_t \Delta t^{\frac{3}{2}}$$

Both Δt terms have exponents greater than one and are correctly categorized as higher order. The term in blue in Equation 62, on the other hand, is

$$\begin{aligned}(\Delta y)^2 &= \left(\mu \Delta t + \sigma \varepsilon_t \sqrt{\Delta t}\right)^2 \\ &= \mu^2 (\Delta t)^2 + \sigma^2 \varepsilon_t^2 \Delta t + 2\mu\sigma\varepsilon_t \Delta t^{\frac{3}{2}} \\ &= \sigma^2 \varepsilon_t^2 \Delta t + H(\Delta t)\end{aligned} \tag{63}$$

$H(\Delta t)$ indicates higher order terms of Δt, but $\sigma^2 \varepsilon_t^2 \Delta t$ does not qualify as higher order. It's clear that the blue term from Equation 62 actually includes a non-higher order term that must be dealt with (specifically, the green term in Equation 63). Taking the expectation of the green term from Equation 63 gives

$$E\left(\sigma^2 \varepsilon_t^2 \Delta t\right) = \sigma^2 E\left(\varepsilon_t^2\right) \Delta t = \sigma^2 \Delta t \tag{64}$$

and the variance is

$$\begin{aligned}\mathrm{Var}\left(\sigma^2 \varepsilon_t^2 \Delta t\right) &= E\left(\sigma^4 \varepsilon_t^4 \Delta t^2\right) - E\left(\sigma^2 \varepsilon_t^2 \Delta t\right)^2 \\ &= E\left(\varepsilon_t^4\right)\sigma^4 \Delta t^2 - \sigma^4 E\left(\varepsilon_t^2\right)^2 \Delta t^2 \\ &= 3\sigma^4 \Delta t^2 - \sigma^4 \Delta t^2 = 2\sigma^4 \Delta t^2\end{aligned}$$

since $E\left(\varepsilon_t^4\right) = 3$ when ε_t is a standard normal random variable.[12] Following from Equations 63 and 64, $(\Delta y)^2 \to \sigma^2 dt$ as $\Delta t \to 0$. Thus, the stochastic term $(\Delta y)^2$ approaches a non-stochastic term in the limit. We can then replace $(\Delta y)^2$ with $\sigma^2 dt$ in the Taylor expansion of ΔG from Equation 62, giving us

$$dG = \frac{\partial G}{\partial y} dy + \frac{\partial G}{\partial t} dt + \frac{1}{2}\frac{\partial^2 G}{\partial y^2}\sigma^2 dt \tag{65}$$

Recall from Equation 58 that $dy = \mu dt + \sigma dW_t$. Plugging this into Equation 65 gives us

$$dG = \frac{\partial G}{\partial y}(\mu dt + \sigma dW_t) + \frac{\partial G}{\partial t}dt + \frac{1}{2}\frac{\partial^2 G}{\partial y^2}\sigma^2 dt$$

$$= \frac{\partial G}{\partial y}\mu dt + \frac{\partial G}{\partial y}\sigma dW_t + \frac{\partial G}{\partial t}dt + \frac{1}{2}\frac{\partial^2 G}{\partial y^2}\sigma^2 dt$$

$$= \left(\frac{\partial G}{\partial y}\mu + \frac{\partial G}{\partial t} + \frac{1}{2}\frac{\partial^2 G}{\partial y^2}\sigma^2\right)dt + \frac{\partial G}{\partial y}\sigma dW_t$$

This is Ito's lemma applied to derivative $G(y)$, when y follows a generalized Wiener process! If y follows the process $dy = \mu dt + \sigma dW_t$, then $G(y,t)$ follows the process

$$dG = \left(\frac{\partial G}{\partial y}\mu + \frac{\partial G}{\partial t} + \frac{1}{2}\frac{\partial^2 G}{\partial y^2}\sigma^2\right)dt + \frac{\partial G}{\partial y}\sigma dW_t \tag{66}$$

CONCEPT REFRESHER 4.3: HIGHER ORDER TERMS

To understand which terms qualify as higher order, we introduce the notation $o(h)$. A function f is said to be $o(h)$ if $\lim_{h\to 0}\frac{f(h)}{h} = 0$. Effectively, if a term is $o(h)$ it can be considered higher order and ignored under appropriate circumstances.

Consider $f(h) = h$ and $g(h) = h^2$. Checking f and g, we have

$$\lim_{h\to 0}\frac{f(h)}{h} = \lim_{h\to 0}\frac{h}{h} = 1$$

$$\lim_{h\to 0}\frac{g(h)}{h} = \lim_{h\to 0}\frac{h^2}{h} = \lim_{h\to 0} h = 0$$

Accordingly, we can say that g is $o(h)$ and f is *not* $o(h)$. This means g is treated as a higher order term and may be ignored under appropriate circumstances. f would not be treated as a higher order term.

Note that if $\lim_{h\to 0}\frac{f(h)}{h} < \infty$, we would say f is $O(h)$ since f is finite. It is only when the limit of the function approaches zero that we say the function is $o(h)$, and hence higher order.

4.2.4 Geometric Brownian Motion

A perceived problem with the generalized Wiener process $dy = \mu dt + \sigma dW_t$ from Equation 58 is that the magnitude of the drift and volatility terms is independent of y. It has been observed that stock prices tend to exhibit drift and volatility proportional to the price of the stock. In other words, as y increases the deviations in Δy tend to increase, and vice-versa for y decreasing. If this weren't the case, then as a stock price increased, the growth and volatility would become smaller and smaller relative to the price.

To accommodate proportionality in drift and volatility, the generalized Wiener process needs to be adjusted. Rewriting the process, we have[13]

$$dS_t = \mu S_t dt + \sigma S_t dW_t \tag{67}$$

We've switched out y_t for S_t to be clear that we're now specifically talking about a process for the price S_t of a stock. Equation 67 is referred to as geometric Brownian motion (GBM), which is simply a special type of Ito process where μ and σ are assumed constant. By multiplying the drift and variance terms by S_t, we ensure that the drift and variance rates will stay proportional to the stock price as it rises and falls.

Further, we know that a stock price can't go below zero. Accordingly, we need to be sure that the GBM model does not allow S_t to fall below zero. To do this, we must first decide how S_t is distributed. We know it can't be normally distributed, since this would imply S_t can be negative. Instead, we assume S_t is lognormally distributed, which means the natural logarithm of S_t is normally distributed. This assumption is convenient because a lognormal random variable cannot fall below zero.

However, we must now ask what process the log of the stock price follows. We can refer to the log of the stock price as $G(S_t, t) = \ln(S_t)$. We know the process for the price S_t (from Equation 67), but $G(S_t, t)$ represents a functional transformation. Fortunately, we just learned Ito's lemma, allowing us to derive the process for the function of a process. Ito's lemma says

$$dG = \left(\frac{\partial G}{\partial S_t} \mu S_t + \frac{\partial G}{\partial t} + \frac{1}{2} \frac{\partial^2 G}{\partial S_t^{\,2}} \sigma^2 S_t^2 \right) dt + \frac{\partial G}{\partial S_t} \sigma S_t dW_t \tag{68}$$

Note that this form is slightly different from that of Equation 66 since it is now applied to the GBM process rather than the generalized Wiener process. Calculating each term for $G(S_t, t) = \ln(S_t)$, we have

$$\frac{\partial G}{\partial S_t} = \frac{1}{S_t}; \quad \frac{\partial G}{\partial t} = 0; \quad \frac{1}{2}\frac{\partial^2 G}{\partial S_t^2} = \frac{1}{2}\frac{\partial G}{\partial S_t}\left(\frac{1}{S_t}\right) = \frac{1}{2}\left(\frac{-1}{S_t^2}\right)$$

Plugging these values into Ito's lemma from Equation 68 gives

$$dG = d\ln(S_t) = \left(\frac{1}{S_t}\mu S_t + \frac{1}{2}\frac{-1}{S_t^2}\sigma^2 S_t^2\right)dt + \frac{1}{S_t}\sigma S_t dW_t$$

$$d\ln(S_t) = \left(\mu - \frac{\sigma^2}{2}\right)dt + \sigma dW_t \tag{69}$$

Relative to the process for S_t, the process for $G(S_t, t)$ has a different drift rate and the proportionality constant has fallen out (i.e., S_t no longer shows up in either the drift or noise terms). In words, the log price of a stock follows a generalized Wiener process with drift rate $\mu - \frac{\sigma^2}{2}$ and variance rate σ^2.

Before continuing, let's solve this stochastic differential equation for S_t. Restating Equation 69 in discrete terms and rearranging, we get

$$\Delta\ln(S_t) = \ln(S_t) - \ln(S_0) = \left(\mu - \frac{\sigma^2}{2}\right)\Delta t + \sigma\Delta W_t$$

So

$$\ln(S_t) = \ln(S_0) + \left(\mu - \frac{\sigma^2}{2}\right)\Delta t + \sigma\Delta W_t$$

Reversing the logarithm gives us

$$S_t = S_0 \exp\left[\left(\mu - \frac{\sigma^2}{2}\right)\Delta t + \sigma\Delta W_t\right] \tag{70}$$

We have solved the stochastic differential equation and can now confirm that the price cannot become negative (since the exp function cannot output a negative number). We can then say that the log price from time t to time T (where t is today and T is some point in the future) follows a normal distribution with drift $\left(\mu - \frac{\sigma^2}{2}\right)(T-t)$ and variance $\sigma^2(T-t)$:

$$\ln(S_T) - \ln(S_t) \sim N\left[\left(\mu - \frac{\sigma^2}{2}\right)(T-t), \sigma^2(T-t)\right]$$

$$\ln(S_T) \sim N\left[\ln(S_t) + \left(\mu - \frac{\sigma^2}{2}\right)(T-t), \sigma^2(T-t)\right] \tag{71}$$

NOTES

1 This section is an adaptation of multiple sources, including Hull (2018), Gupta (2014), Tsay (2010), and Ross (2019).

2 Independent and stationary increments can be thought of as the continuous time analog of independent and identically distributed ("i.i.d.") random variables in discrete time. In discrete time, identically distributed means realizations of the random variable ε_t are drawn from the same probability distribution, with constant parameters, for all t. Independent means that the noise variable at time t_i has no impact on the noise variable at time t_j for $i \neq j$.

3 For a rigorous proof see Donsker's theorem, which is outside the scope of this book.

4 The function that generates the normal distribution is $\phi(x) = \frac{1}{\sigma\sqrt{2\pi}} \exp\left[-\frac{1}{2}\left(\frac{x-\mu}{\sigma}\right)^2\right]$ where μ is the mean and σ is the standard deviation. The literature may also use $f(x)$ instead of $\phi(x)$. Typically, when we say we "standardize" a variable, we de-mean the variable and then divide by its standard deviation. So you would replace every x in your time series with $x_{\text{standard}} = \frac{x - \text{mean of } X}{\text{Standard deviation of } X}$. It is common to use $Z = \frac{X-\mu}{\sigma}$ when referring to a standardized normal random variable. When X is normally distributed, Z will have a mean and standard deviation of zero and one, respectively. After this transformation of turning X into Z, Z would follow a standard normal distribution. The function that generates the standard normal distribution is $\phi(z) = \frac{1}{\sqrt{2\pi}} \exp\left[-\frac{1}{2}z^2\right]$.

5 Robert Brown was a 19th-century botanist studying pollen. Under a microscope, Brown noticed the pollen exhibited a fidgety motion when suspended in water. Through repeated experimentation, Brown determined it was not purposeful living motion. While he could not explain the motion, this observation is credited as the discovery of the motion process that became known as Brownian motion.

6 The Wiener process is named after mathematician Norbert Wiener, who provided formal mathematical structure to Brownian motion in Wiener (1923).

7 Notice the similarity with Equation 55. They are in fact the same outside of a notation change.

8 Carl Friedrich Gauss is a historically renowned mathematician and physicist from the 19th century with extensive contributions to mathematics.

9 The Ito process (and Ito's lemma) is named after Kiyosi Ito, a 20th-century mathematician who specialized in probability theory. For convenience, we drop the (y_t, t) after each variable hereafter.

10 The use of the term *derivative* here differs from the calculus definition used prior to this point. G being a derivative of y means that G's value is *derived* from the value of y. Derivatives, also known as contingent claims, will be discussed in detail in Section 5.

11 The assumption that higher order terms can be ignored reflects an assumption of normality. Higher moments of a normally distributed random variable are constant (e.g., skew is zero and kurtosis is three). The moments of a distribution are discussed more in Section 6.2.1. If instead one is dealing with fat tailed variables, higher order moments may explode (be infinite), meaning higher order terms cannot be ignored. Accordingly, for Ito's lemma to describe the process for G, we must assume y does not "jump". For a refresher on what defines a higher order term, see Concept Refresher 4.3.

12 $E\left(\varepsilon_t^4\right)$ reflects the kurtosis, also known as the fourth central moment, of ε_t. It is a measure of the thickness of the tails (far left and far right) of the probability distribution. This concept is unimportant for now, other than to note that the kurtosis of a standard normal random variable is three. The moments of a distribution are discussed more in Section 6.2.1.

13 Recall a similar form from Equation 46.

SECTION 5

Derivative Pricing

5.1 NO ARBITRAGE AND RISK-NEUTRAL PROBABILITIES[1]

Before diving into derivative pricing, it is important to give context to concepts that will arise in the coming sections. A derivative, also known as a contingent claim, is a security the price of which is *derived* from some other underlying asset or security. To establish a price for a derivative, we invoke the no-arbitrage principle. That is, the price of two securities with identical cash flows in all possible future states of the world must have the same price today. If this weren't true, then one could purchase and sell the identical assets in a way that would earn an immediate payoff without any risk of loss.

A *replicating portfolio* is a portfolio constructed to replicate the payoff of a derivative security. By the no-arbitrage principle, the price of a derivative should be the same as the price of a replicating portfolio that replicates its payoff. To formally define what it means for an arbitrage opportunity to exist, consider a world in which there are m assets. The prices of these m assets at time 0 are represented by the vector $\boldsymbol{S}_0 = \left[S_0^1, \ldots, S_0^m \right]$. We are interested in representing these asset prices at some point in the future, say at time 1. We know these asset prices will depend on the state of the world at time 1, but there could be an infinite number of these possible future states.

For our purposes, assume there are n possible future states represented by the set $\Omega = \{\omega_1, \ldots, \omega_n\}$, where ω_1 is the first possible state of the world, ω_2 the second possible state, and so on. Picking one particular state, say ω_j, $1 \le j \le n$, we can represent the prices of the m assets at time 1 with the vector $\boldsymbol{S}_1(\omega_j) = \left[S_1^1(\omega_j), \ldots, S_1^m(\omega_j) \right]$.

DOI: 10.1201/9781032687650-5

Next, our goal is to earn an arbitrage profit by constructing a portfolio from the m assets. When constructing a portfolio, one incurs a cost at time 0 and a payoff at time 1. To successfully exploit an arbitrage opportunity, this portfolio must have non-positive cost and non-negative payoff in *all* future states of the world, but must also have a strictly negative cost or generate a strictly positive payoff in *some* future states. Said differently, this portfolio must generate a positive payoff at either time 0 or 1 and have no risk of loss at either time 0 or 1, no matter which state of the world is realized at time 1.

To formalize these requirements for an arbitrage portfolio, consider a portfolio with holdings $\boldsymbol{y} = \left[y_1, \ldots, y_m \right]$, where y_1 is the amount invested in asset 1, and so on. To be an arbitrage portfolio, it must be true that[2]

$$\boldsymbol{S}_0 \boldsymbol{y}^T = S_0^1 y_1 + S_0^2 y_2 + \cdots + S_0^m y_m \leq 0$$

$$\boldsymbol{S}_1 \left(\omega_j \right) \boldsymbol{y}^T = S_1^1 \left(\omega_j \right) y_1 + S_1^2 \left(\omega_j \right) y_2 + \cdots + S_1^m \left(\omega_j \right) y_m \geq 0, \quad j = 1, \ldots, n$$

and at least one of the inequalities is strict (i.e., either at time 0 or for one of the $j = 1, \ldots, n$ at time 1).

Now that we understand what it means for arbitrage to exist, we next define a *positive linear pricing rule* (also known as a *state price deflator*). Denote $\boldsymbol{x} = \left[x_1, \ldots, x_n \right]$ a set of positive numbers. For $\boldsymbol{x}$ to be a positive linear pricing rule, it must satisfy the identity in Equation 72:

$$\boldsymbol{S}_0 = \sum_{j=1}^{n} \boldsymbol{S}_1 \left(\omega_j \right) x_j \tag{72}$$

In words, the vector of prices at time 0 must reflect a weighted average of the prices at time 1 over all n possible states. The positive linear pricing rule $\boldsymbol{x}$ acts as the vector of weights. The existence of a positive linear pricing rule is both necessary and sufficient to ensure that no arbitrage opportunities exist. Notice that $\boldsymbol{x}$ must simultaneously capture two distinct forces: discounting and probability, that is, x_j accounts for the probability that state ω_j occurs *and* accounts for discounting from time 1 back to time 0.

Consider now a special asset known as the *numeraire*. The numeraire may be thought of as a unit of account for asset prices. In monetary economies, it is common for money to play the role of numeraire, though other assets such as commodities and financial securities can also play this role.

In the current analysis, a risk-free security will be used as the numeraire. As we will see, this choice of numeraire will act as a bridge through time. By expressing asset prices in terms of the numeraire, the time value of money will be implicitly accounted for.

Formally, define a numeraire as an asset with a strictly positive payoff in all possible future states of the world, that is, the price of the numeraire at time 1 is $S_1^{rf}(w_j) > 0$ for $j = 1,\ldots,n$. The superscript rf is used to denote the numeraire price because of the use of a risk-free security as the numeraire. Specifically, we assume that the numeraire is a risk-free asset with return r such that $S_0^{rf} = 1$ is its price at time 0 and $S_1^{rf}(\omega_j) = 1 + r$ is its price at time 1. Notice that the price of the numeraire at time 1 is independent of the state of the world.

S_0^i denotes the price of some risky asset i at time 0 and $S_1^i(\omega_j)$ denotes the price of the same risky asset i at time 1 for state ω_j. From Equation 72, we know the price of the risky asset and the price of the numeraire can be expressed, respectively, as

$$
\begin{aligned}
S_0^i &= \sum_{j=1}^{n} S_1^i(\omega_j) x_j \\
S_0^{rf} &= \sum_{j=1}^{n} S_1^{rf}(\omega_j) x_j
\end{aligned}
\tag{73}
$$

Using Equation 73 and the fact that $S_1^{rf}(\omega_j) = 1 + r, \forall j$, the price of risky asset i relative to the numeraire is

$$
\begin{aligned}
\frac{S_0^i}{S_0^{rf}} &= \frac{\sum_{j=1}^{n} S_1^i(\omega_j) x_j}{S_0^{rf}} \\
&= \sum_{j=1}^{n} \frac{S_1^i(\omega_j) x_j}{S_0^{rf}} \frac{S_1^{rf}(\omega_j)}{S_1^{rf}(\omega_j)} \\
&= \sum_{j=1}^{n} \frac{S_1^i(\omega_j)}{S_1^{rf}(\omega_j)} \frac{S_1^{rf}(\omega_j) x_j}{S_0^{rf}} = E_{\mathbb{Q}}\left(\frac{S_1^i}{S_1^{rf}}\right)
\end{aligned}
\tag{74}
$$

$$\frac{S_0^i}{S_0^{rf}} = \sum_{j=1}^{n} \frac{S_1^i(\omega_j)}{S_1^{rf}(\omega_j)} \frac{S_1^{rf}(\omega_j)x_j}{\sum_{j=1}^{n} S_1^{rf}(\omega_j)x_j}$$

$$= \sum_{j=1}^{n} \frac{S_1^i(\omega_j)}{1+r} \frac{(1+r)x_j}{(1+r)\sum_{j=1}^{n} x_j} \tag{75}$$

$$= \frac{1}{1+r} \sum_{j=1}^{n} S_1^i(\omega_j) \frac{x_j}{\sum_{j=1}^{n} x_j} = \frac{1}{1+r} E_{\mathbb{Q}}(S_1^i)$$

Denote $\mathbb{Q}(\omega_j) = \frac{S_1^{rf}(\omega_j)x_j}{S_0^{rf}} = \frac{(1+r)x_j}{(1+r)\sum_{j=1}^{n} x_j} = \frac{x_j}{\sum_{j=1}^{n} x_j}$ the *risk-neutral probability measure* and note that

$$\sum_{j=1}^{n} \mathbb{Q}(\omega_j) = \sum_{j=1}^{n} \frac{S_1^{rf}(\omega_j)x_j}{S_0^{rf}} = \frac{1}{S_0^{rf}} \sum_{j=1}^{n} S_1^{rf}(\omega_j)x_j = \frac{1}{S_0^{rf}} S_0^{rf} = 1 \tag{76}$$

It follows from Equation 76 that $\mathbb{Q}$ is a probability mass function. The relative price of risky asset i is shown in Equation 74 to equal the expected relative payoff, where expectation is taken with respect to the probability measure $\mathbb{Q}$.[3] Notice also in Equation 75 that we have successfully decoupled the time discounting component from the state probabilities. Discounting now occurs at the risk-free rate, while $\mathbb{Q}(\omega_j)$ acts only as a state probability measure.

It follows that a positive linear pricing rule has two alternative representations:

- Stochastic discount factor
- Risk-neutral probability measure

In both of these alternative representations, the set of future states $\Omega = \{\omega_1, \ldots, \omega_n\}$ is interpreted as a probability space. Assume that the space Ω has probability measure $\mathbb{P}$. Think of $\mathbb{P}$ as the real-world probability that some state of the world will occur. The payoff at time 1 for asset $i, i = 1, \ldots, m$ may now be viewed as a random variable $S^i : \Omega \to \mathbb{R}$. In words, the random variable S^i takes some state of the world from the set Ω as an input and outputs a real number from the set of real numbers $\mathbb{R}$. The output is the return for asset i from time 0 to 1.

A *stochastic discount factor D* is also a random variable, instead defined as $D:\Omega \to \mathbb{R}$, such that the identity in Equation 77 holds

$$\boldsymbol{S}_0 = E_{\mathbb{P}}\left(D\boldsymbol{S}_1\right) = \sum_{j=1}^{n} \boldsymbol{S}_1\left(\omega_j\right)D\left(\omega_j\right)\mathbb{P}\left(\omega_j\right) \tag{77}$$

where $D\left(\omega_j\right)$ is the stochastic discount factor associated with the state of the world ω_j, $\mathbb{P}\left(\omega_j\right)$ is the real-world probability of state ω_j, and $E_{\mathbb{P}}$ is the expectation with respect to $\mathbb{P}$. In this representation, the positive linear pricing rule is defined as $x_j = D\left(\omega_j\right)\mathbb{P}\left(\omega_j\right)$. One could equivalently write Equation 77 as

$$\boldsymbol{S}_0 = E_{\mathbb{P}}\left(D\boldsymbol{S}_1\right) = \sum_{j=1}^{n} \frac{\boldsymbol{S}_1\left(\omega_j\right)}{1+\tilde{r}_j}\mathbb{P}\left(\omega_j\right)$$

where $\tilde{r}_j$ is the risky discount rate in state ω_j.

A risk-neutral probability measure is a probability measure $\mathbb{Q}$ for the space Ω such that the identity in Equation 78 holds

$$\boldsymbol{S}_0 = \frac{1}{1+r}E_{\mathbb{Q}}\left(\boldsymbol{S}_1\right) = \frac{1}{1+r}\sum_{j=1}^{n} \boldsymbol{S}_1\left(\omega_j\right)\mathbb{Q}\left(\omega_j\right) \tag{78}$$

where $E_{\mathbb{Q}}$ denotes the expectation with respect to the risk-neutral probability measure $\mathbb{Q}$. In this representation, the positive linear pricing rule is defined as $x_j = \frac{1}{1+r}\mathbb{Q}\left(\omega_j\right)$.

Restating both positive linear pricing rule representations, we have (Table 5.1):

Recalling that $\frac{1}{1+r}$ is the risk-free discount factor and $\frac{1}{1+\tilde{r}_j}$ the risky discount factor, the stochastic discount factor approach can be said to incorporate risk through the use of risky discount rates and real-world probabilities. That is, risk is incorporated via the discount rate. The risk-neutral probability measure approach instead discounts by the risk-free rate and accounts for risk by adjusting the probability measure from $\mathbb{P}$ to $\mathbb{Q}$.[4]

TABLE 5.1 Positive Linear Pricing Rules

Rule	x_j
Stochastic discount factor	$\frac{1}{1+\tilde{r}_j}\mathbb{P}(\omega_j)$
Risk-neutral probability measure	$\frac{1}{1+r}\mathbb{Q}(\omega_j)$

To summarize the above, the existence of a positive linear pricing rule implies there are both a risk-neutral probability measure and a positive stochastic discount factor. Each of these is equivalent to saying there are no arbitrage opportunities. This is known as the *fundamental theorem of asset pricing*. Of course, arbitrage opportunities *do* periodically present themselves in the real world. When this happens, the identities from Equations 72–78 can be said to no longer hold as long as the arbitrage opportunity remains.

A basic application of these concepts will help motivate methods introduced later in this section. Consider once again a world previously described, but now set $m=2$ and $n=2$. That is, the world now has only $m=2$ assets: a risk-free asset and a risky asset. The risk-free rate is denoted as r and the price of the risky asset at time 0 is denoted as S_0. In addition, there are only $n=2$ possible future states at time 1. Denote these states H and T. Formally, $\Omega=\{H,L\}$. As before, the price of the risky asset at time 1 is either $S_1(H)$ or $S_1(L)$. Think of H and L as denoting "high" and "low" scenarios, respectively, meaning the price of the risky asset is higher in the high scenario than it is in the low scenario.

We can define $S_1(H)=US_0$ and $S_1(L)=DS_0$, $0<D<U$. Here D and U are scalars that multiply the price at time 0 to arrive at the price at time 1. Since $U>D$, we ensure that $S_1(H)>S_1(L)$. Solving for the risk-neutral probability $\mathbb{Q}$, we have

$$S_0=\frac{1}{1+r}\left[S_1(H)\mathbb{Q}(H)+S_1(L)\mathbb{Q}(L)\right]$$

$$S_0=\frac{1}{1+r}\left[S_0U\mathbb{Q}(H)+S_0D\mathbb{Q}(L)\right]$$

$$S_0=\frac{S_0}{1+r}\left[U\mathbb{Q}(H)+D\mathbb{Q}(L)\right]$$

$$1+r = U\mathbb{Q}(H)+D\mathbb{Q}(L)$$

$$1+r = U\mathbb{Q}(H)+D(1-\mathbb{Q}(H))$$

$$1+r = U\mathbb{Q}(H)+D-D\mathbb{Q}(H)$$

$$1+r-D = \mathbb{Q}(H)(U-D)$$

$$\mathbb{Q}(H) = \frac{1+r-D}{U-D} \tag{79}$$

For there to be no arbitrage in this world, the risk-neutral probability must be a valid probability mass function. Specifically, we must have $D < 1+r < U$ (to avoid negative probabilities) and $\mathbb{Q}(H)+\mathbb{Q}(L)=1$. These concepts will emerge again in Sections 5.3.1 and 5.3.2 when deriving a closed-form equation useful for pricing derivatives.

5.2 BLACK–SCHOLES–MERTON DIFFERENTIAL EQUATION[5]

We now know the process for both the stock price S (GBM) and, using Ito's lemma, the process for a derivative G of the stock price.[6] The derivative we primarily work with in this section is a European call option.

CONCEPT REFRESHER 5.1: OPTION CONTRACTS

An option is a derivative. There are two basic types of options: calls and puts. Calls give the owner the right to purchase a given stock at the strike price, while a put gives the owner the right to sell a given stock at the strike price. The strike price is a contractual value determined prior to purchase of the option.

There are many different styles of options. European options may only be exercised on the expiration date, while American options may be exercised any time on or prior to the expiration date.

To visualize calls vs. puts, consider their payoff charts. The first chart in Figure 5.1 shows the payoff for a call option with a strike of \$12 per share and a premium of \$1. Upon exercise, the option will have value if the underlying stock price is greater than the strike price. Notice the kink in the payoff at the strike price. Below the strike price, the investor has a payoff of –\$1, reflecting the cost of the premium.

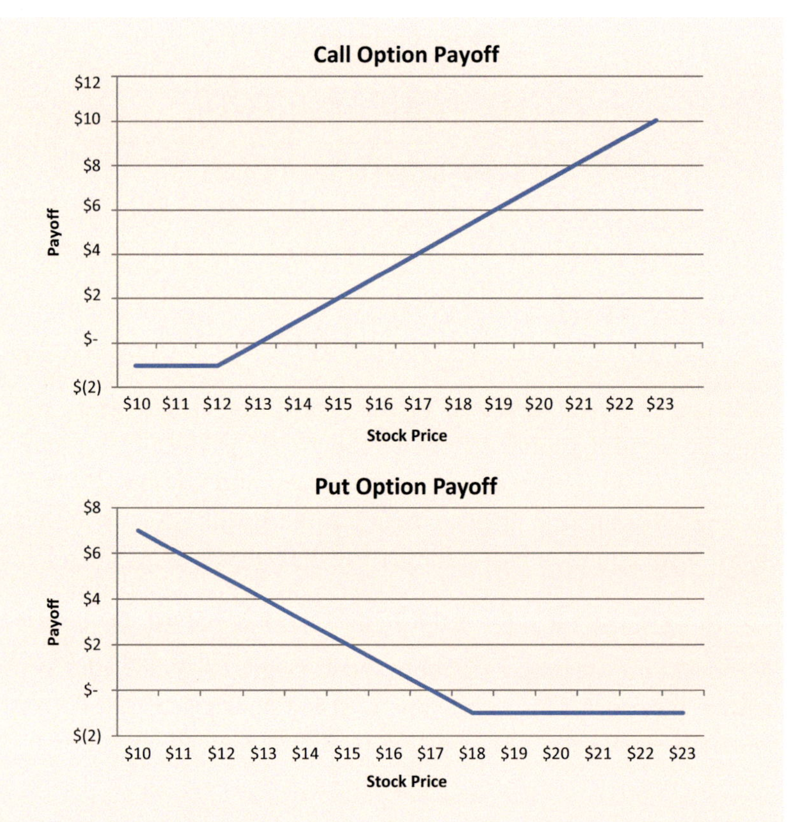

FIGURE 5.1 Option payoffs.

The second chart in Figure 5.1 shows the payoff for a put option with a strike of $18 per share and a premium of $1. The put option payoff has the reverse structure. Above the strike price, the investor has a payoff of −$1, reflecting the cost of the premium.

In practice, equity option contracts are generally for 100 shares of the underlying stock, meaning the payoff must be multiplied by 100.

The payoff charts from Figure 5.1 indicate *intrinsic value*, which is the only source of value when exercising an option. Intrinsic value is defined as the stock price less the strike price for a call option and the strike price less the stock price for a put option. If intrinsic value is positive when the option is exercised, the option is said to be *in-the-money*. Prior to exercise, options have both intrinsic value and *time value*. Time value is the portion of the option price exceeding that of intrinsic value. If an option is *out-of-the-money*, meaning the stock price is below the strike price for a call or above the strike price for a put, then the entire value of the option is attributed to time value. Note that the premium paid *is* the price of the option at initiation of the contract.

Restating the discretized version of GBM and Ito's lemma for GBM, we have[7]

$$\Delta S_t = \mu S_t \Delta t + S_t \sigma \Delta W_t \tag{80}$$

$$\Delta G_t = \left(\frac{\partial G_t}{\partial S_t} S_t \mu + \frac{\partial G_t}{\partial t} + \frac{1}{2} \frac{\partial^2 G_t}{\partial S_t^2} S_t^2 \sigma^2 \right) \Delta t + \frac{\partial G_t}{\partial S_t} S_t \sigma \Delta W_t \tag{81}$$

The stochastic term ΔW_t is tricky, and we would like to get rid of it. Since ΔW_t is present in both of these processes, it makes sense to construct a portfolio consisting of both the underlying stock (which follows the ΔS_t process) and the call option (which follows the ΔG_t process) in such a way that ΔW_t disappears. We can do this by going short[8] the call option and going long (purchasing) the stock. The question is, how much of each?

To answer this question, we must introduce the *delta*. Delta is one of many Greek terms, each of which is used to measure the change in the price of an option given a change in some variable. Delta specifically refers to the change in the price G_t of an option given an infinitesimally small change in the price S_t of the underlying security.

Consider first the change in G_t given a larger change in S_t. This can be written as $\frac{\Delta G_t}{\Delta S_t}$. Loosely speaking, if a \$10 change in Apple's stock price causes the price of a derivative on Apple's stock to change by \$6, then the delta is $\frac{\Delta G_t}{\Delta S_t} = \frac{6}{10} = 0.6$. Allowing ΔS_t to approach zero, we have $\lim_{\Delta S_t \to 0} \frac{\Delta G_t}{\Delta S_t} = \frac{\partial G_t}{\partial S_t}$. This is the delta of the derivative G. We will use the symbol $\delta = \frac{\partial G_t}{\partial S_t}$ to represent delta.

For infinitesimally small changes in the stock price, if you own exactly δ shares of stock and are short one call option, the value of the combined portfolio should not change at all. The change in the value of the call option position is exactly offset by the change in the value of the stock position. Since this portfolio's value is insulated to changes in the underlying stock price, this portfolio can be thought of as risk-free. This is referred to as a *delta-hedged* portfolio.

Before describing this formally, consider an important assumption here: to remain risk-free, this portfolio must be rebalanced *dynamically*. As S_t changes, so too will δ. As such, the portfolio is only hedged for an infinitesimally small amount of time. In other words, the amount of stock owned must always be equal to δ, even as δ changes for every infinitesimal change in S_t. If this dynamic hedge is not maintained, then this portfolio would not be risk-free. As we'll see, this would complicate the drift of the portfolio.

To better understand this assumption, consider Figure 5.2. This illustration is meant only to demonstrate the concept of dynamic hedging, so

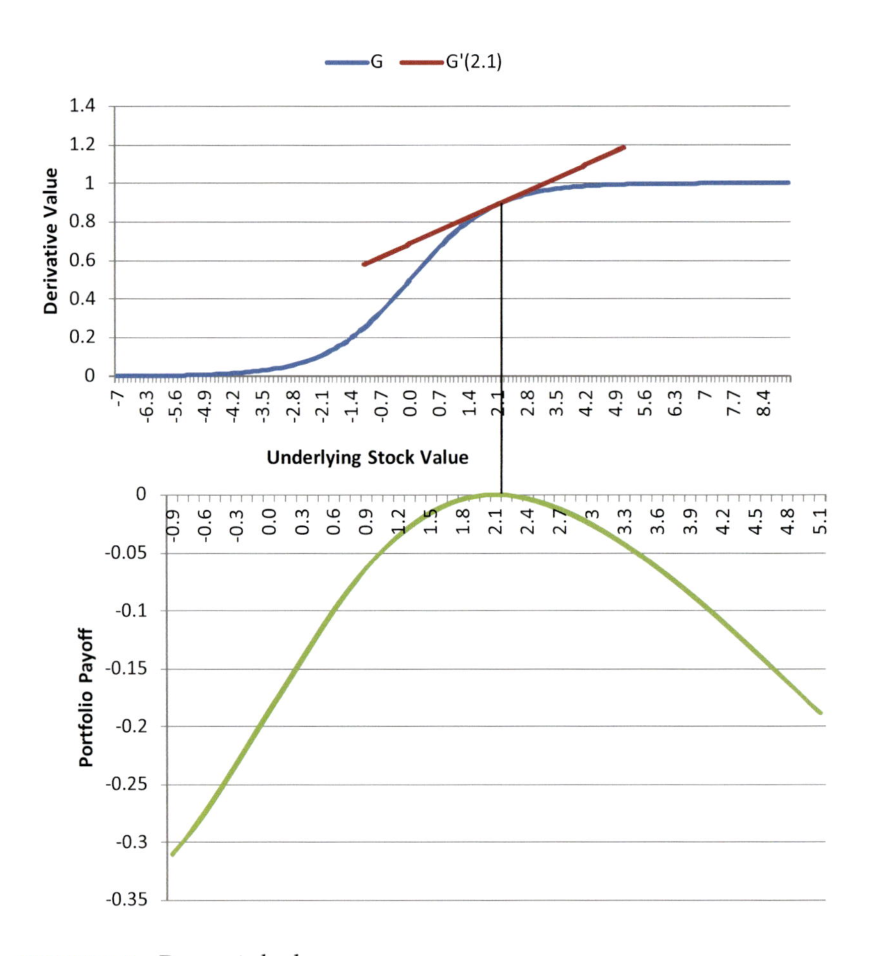

FIGURE 5.2 Dynamic hedge.

the values themselves are meaningless. In the upper graph, we have the blue line representing the value G of a derivative for different values S of the underlying stock across the x-axis. The value $S = 2.1$ is arbitrarily singled out. The red line represents the linear approximation of G. Taking the derivative $\frac{\partial G}{\partial S}$ at $G(2.1)$ gives us the delta $G'(2.1) = \delta$. If we were to zoom in to an infinitesimally small level, we would find that the red and blue lines are perfectly parallel at $S = 2.1$.

Say a portfolio consisted of two positions, one long and one short, and these two position values were represented by the red and blue lines. Then for an infinitesimally small change in S, the portfolio would not experience any change in value. This is demonstrated in the lower graph in Figure 5.2, which depicts the payoff of this fictitious portfolio for different values of S. At $S = 2.1$, if S changes by an infinitely small amount, the value of the portfolio does not change. As S changes, however, the red line separates from the blue line, representing an imperfect hedge. In such a case, the portfolio value would change as S continues to move. To maintain a constant value for this portfolio, the hedge would have to be rebalanced every time S changes by more than an infinitely small amount.

This visualization demonstrates the importance of the dynamic hedge assumption. Without it, we could not assume the hedged portfolio is risk-free.[9]

Describing this dynamically-hedged portfolio formally, we have

$$V_t = -G_t + \delta S_t \tag{82}$$

where V_t is the value of the portfolio at time t. The negative sign in front of G_t represents the fact that we are short the call option, while the coefficient $\delta = \frac{\partial G_t}{\partial S_t}$ signifies that we are long δ number of shares of the underlying stock S_t. The change in the value of the portfolio is then

$$\Delta V_t = -\Delta G_t + \delta \Delta S_t \tag{83}$$

Plugging in the processes followed by ΔG_t and ΔS_t from Equations 80 and 81, and remembering once more that $\delta = \frac{\partial G_t}{\partial S_t}$, we can write Equation 83 as

$$\Delta V_t = -\Delta G_t + \delta \Delta S_t$$

$$= -\left[\left(\frac{\partial G_t}{\partial S_t} S_t \mu + \frac{\partial G_t}{\partial t} + \frac{1}{2}\frac{\partial^2 G_t}{\partial S_t^2} S_t^2 \sigma^2\right)\Delta t + \frac{\partial G_t}{\partial S_t} S_t \sigma \Delta W_t\right] + \delta\left[\mu S_t \Delta t + S_t \sigma \Delta W_t\right]$$

$$= \left(-\delta S_t \mu - \frac{\partial G_t}{\partial t} - \frac{1}{2}\frac{\partial^2 G_t}{\partial S_t^2} S_t^2 \sigma^2\right)\Delta t - \delta S_t \sigma \Delta W_t + \delta\ \mu S_t \Delta t + \delta S_t \sigma \Delta W_t$$

$$= \left(-\delta S_t \mu - \frac{\partial G_t}{\partial t} - \frac{1}{2}\frac{\partial^2 G_t}{\partial S_t^2} S_t^2 \sigma^2\right)\Delta t + \delta\ \mu S_t \Delta t$$

$$= -\delta S_t \mu \Delta t - \frac{\partial G_t}{\partial t}\Delta t - \frac{1}{2}\frac{\partial^2 G_t}{\partial S_t^2} S_t^2 \sigma^2 \Delta t + \delta\ \mu S_t \Delta t$$

$$= \left(-\frac{\partial G_t}{\partial t} - \frac{1}{2}\frac{\partial^2 G_t}{\partial S_t^2} S_t^2 \sigma^2\right)\Delta t \tag{84}$$

Notice that the stochastic terms containing the Wiener process ΔW_t have disappeared! Under the assumption that no arbitrage opportunities exist,[10] the portfolio should be expected to earn the risk-free rate in the time interval Δt. We will refer to the risk-free rate as r.

Formally, this means

$$\Delta V_t = V_t r \Delta t \tag{85}$$

Equation 85 simply says that the return on the portfolio is equal to the value of the portfolio multiplied by the risk-free rate, and then scaled depending on the holding period Δt.

Let's take a moment to rationalize. We saw in Figure 5.2 that this hedged portfolio should have a payoff of zero in response to an infinitesimally small change in S. Here we're saying the portfolio should have payoff r. This follows from the existence of a risk-free bond. Consider the construction of this hedged portfolio in reverse. Imagine shorting δS amount of stock, in turn, receiving δS amount of proceeds. From these proceeds, you purchase an option at price G. The remaining proceeds equals $\delta S - G$. This is exactly the value V of the portfolio from Equation 82. By investing these proceeds into a risk-free bond, one would generate return r. So while the return of the hedged portfolio for an infinitesimally small change in S is zero, the payoff for a change in *time* is r.[11]

We now have two representations of the return on this risk-free portfolio (Equations 84 and 85). Equating them gives us

$$\left(-\frac{\partial G_t}{\partial t}-\frac{1}{2}\frac{\partial^2 G_t}{\partial S_t^2}S_t^2\sigma^2\right)\Delta t = V_t r\Delta t \tag{86}$$

Plugging in Equation 82 for V_t, Equation 86 becomes

$$\left(-\frac{\partial G_t}{\partial t}-\frac{1}{2}\frac{\partial^2 G_t}{\partial S_t^2}S_t^2\sigma^2\right)\Delta t = \left(-G_t+\delta S_t\right)r\Delta t$$

$$-\frac{\partial G_t}{\partial t}-\frac{1}{2}\frac{\partial^2 G_t}{\partial S_t^2}S_t^2\sigma^2 = -G_t r+\delta S_t r$$

$$\frac{\partial G_t}{\partial t}+\frac{1}{2}\frac{\partial^2 G_t}{\partial S_t^2}S_t^2\sigma^2 = G_t r-\delta S_t r$$

$$\delta S_t r+\frac{\partial G_t}{\partial t}+\frac{1}{2}\frac{\partial^2 G_t}{\partial S_t^2}S_t^2\sigma^2 = G_t r \tag{87}$$

Here we have the Black–Scholes–Merton (BSM) differential equation for pricing derivatives. It states that the risk-free rate of return on the derivative G_t is a function of its delta, rho, and gamma.[12] Assuming the BSM assumptions hold, and that the underlying stock pays no dividends, this identity must be satisfied for a pricing formula to imply no arbitrage opportunities exist.

CONCEPT REFRESHER 5.2: DIVIDENDS

Dividends are cash flows paid from companies to shareholders. They are effectively a partial liquidation of a company. Whether paid from retained earnings or funded by debt, dividends reduce the value of a firm.

The date of record is the day on which companies verify ownership of stock for issuance of dividends. One business day prior is the ex-dividend date. Owners of stock on the ex-dividend date are guaranteed to receive the dividend. As such, the stock price will decline by the exact per share amount of the dividend on the ex-dividend date (though trading activity still determines total price change). This ex-dividend price adjustment is automatic. Mechanically, it is applied to the prior day's closing price and the bids and asks in the order book.

This adjustment to stock prices for dividends cannot be captured by a model, given that the decision to issue a dividend and the amount of the dividend are arbitrary. Accordingly, one must assume that dividends are not paid for the BSM differential equation to hold.

The solution to Equation 87 for any particular derivative depends on boundary conditions. Boundary conditions describe the maximum and minimum values of a security at a certain point in time. For example, for a European call, the boundary condition is

$$G_T = \max(S_T - K, 0) \tag{88}$$

where S_T is the price of the underlying security at time T (expiration), K is the strike price, and G_T is the value of the option at expiration.

Note that this condition is binding only at expiration for European options. Prior to expiration (i.e., at time $t < T$), the price of a European call option is

$$E\left[\max(S_T - K, 0)\right]e^{-\alpha(T-t)} \tag{89}$$

The discount rate α is challenging. Each market participant may have their own opportunity cost of capital, meaning no single price emerges. As we will see in Section 5.3.1, this challenge is solved by operating within the risk-neutral world.

5.3 THE BLACK–SCHOLES–MERTON PRICING FORMULA[13]

We now have most of the tools necessary to derive the well-known BSM option pricing formula. Let's start by presenting the end product. The closed-form solution[14] for the price of a European call option, assuming the BSM assumptions hold true, is

$$c_t = S_t\Phi(d_1) - K\Phi(d_2)e^{-r(T-t)}$$

where c_t is the price of the European call option at time t, S_t is the price of the underlying stock, r is the constant risk-free rate, $(T-t)$ is the time to expiration, and K is the strike price. The terms $\Phi(d_1)$ and $\Phi(d_2)$, sometimes written $N(d_1)$ and $N(d_2)$, reference the cumulative probabilities associated with the standard normal values d_1 and d_2:

$$d_1 = \frac{\ln\left(\frac{S_t}{K}\right) + \left(r + \frac{\sigma^2}{2}\right)(T-t)}{\sigma\sqrt{T-t}} \tag{90}$$

$$d_2 = \frac{\ln\left(\frac{S_t}{K}\right) + \left(r - \frac{\sigma^2}{2}\right)(T-t)}{\sigma\sqrt{T-t}} = d_1 - \sigma\sqrt{T-t} \tag{91}$$

σ^2 is the constant variance rate. The red coloring details the only difference between the two formulas. To better understand the Φ terms, recall from Endnote 4 in Section 4 that the PDF for the normal distribution is

$$\phi(x) = \frac{1}{\sigma\sqrt{2\pi}} \exp\left[-\frac{1}{2}\left(\frac{x-\mu}{\sigma}\right)^2\right] \tag{92}$$

Say we define the random variable $Z = \frac{X-\mu}{\sigma}$. If X is normally distributed, then Z will have a standard normal distribution, meaning normally distributed with mean $\mu = 0$ and standard deviation $\sigma = 1$; formally $Z \sim N(0,1)$. When the variable in question is standard normal, the function in Equation 92 simplifies to

$$\phi(z) = \frac{1}{\sqrt{2\pi}} \exp\left[-\frac{1}{2}z^2\right]$$

Taking the definite integral of this function from $-\infty$ to z, one obtains the CDF for the standard normal distribution.

$$\Phi(z) = \int_{-\infty}^{z} \frac{1}{\sqrt{2\pi}} \exp\left(-\frac{1}{2}u^2\right) du$$

As a reminder from Concept Refresher 4.1, one may think of the PDF $\phi(z)$ as describing the probability that a random variable falls between any two values of z, while the CDF $\Phi(z)$ describes the probability that a random variable is less than or equal to some value z.

Once we have input values for all variables in the d_1 and d_2 equations, we may treat d_1 and d_2 as particular values z in the CDF. This allows us to calculate the probability that a standard normal random variable is less than or equal to d_1 or d_2. Once found, these probabilities are plugged in for $\Phi(d_1)$ and $\Phi(d_2)$.

With the end product understood, we may now return to the beginning to derive it. As it turns out, there are several paths one can take to get to the same end point. To better understand the BSM option pricing formula, we will derive it in two ways: the first is termed the risk-neutral approach (Section 5.3.1), and the second is the binomial tree approach (Section 5.3.2).

5.3.1 Risk-Neutral World

In a risk-neutral world, all investors are assumed risk-neutral. What does this mean? Simply put, investors are indifferent to risk. Accordingly, the expected return on all securities equals the risk-free interest rate r. In such a case, one need only discount cash flows at the risk-free rate to obtain their present value, since no investors demand a premium for risk.

Of course, investors tend to be risk-averse in the real world (i.e., as investors take more risk, they require more return). Invoking the risk-neutral assumption, however, is acceptable from an option pricing standpoint, as an option price calculated in the risk-neutral world ends up being the same in the real world. This follows from the fact that the risk-neutral valuation process does not value options in absolute terms. Rather, the value of an option is calculated *in terms of the price of the underlying stock*. Stock prices already reflect investor preferences, so the formulas relating option price to stock price remain the same no matter the risk preferences of investors. This becomes clear upon visual inspection of the BSM differential equation derived in Equation 87, as it does not include any variables that are impacted by investor risk preferences (i.e., real-world drift μ).

Now, recalling the boundary condition for a European call option from Equation 88, the expected value of a European call option at expiration is

$$E_N\left[\max(S_T - K, 0)\right]$$

where the subscript N is used to signify that the expectation is taken in the risk-neutral world. Prior to expiration (i.e., at time $t < T$), the price of a European call option is

$$c_t = E_N\left[\max(S_T - K, 0)\right]e^{-r(T-t)} \tag{93}$$

The second term here discounts the expected value from time T back to time t. Note that r is used in place of α from Equation 89, as we are operating in a risk-neutral world. Let's explore other ways to express $E_N\left[\max(S_T - K, 0)\right]$. Imagine instead that the max function is absent. We would have

$$\begin{aligned} c_t &= E_N\left[S_T - K\right]e^{-r(T-t)} \\ &= \left[E(S_T) - K\right]e^{-r(T-t)} \end{aligned} \tag{94}$$

Without the max function, the optionality has been stripped out. Since expected return μ is equal to the risk-free rate r in a risk-neutral world,

$$E(S_T) = \int_{-\infty}^{\infty} s_T \phi(s_T)\, ds_T = S_0 e^{rT} \tag{95}$$

where $\phi(\cdot)$ is the probability distribution for S_T (which, as previously mentioned, is assumed to be lognormal), and s_T is a realization of S_T. Notice this matches the deterministic future value from Equation 44 in Section 4.1. Plugging Equation 95 into Equation 94 and noting that $(T-t)=T$ when at time $t=0$, we get

$$\begin{aligned}\left[E(S_T)-K\right]e^{-r(T-t)} &= \left[S_0e^{rT}-K\right]e^{-rT}\\ &= S_0e^{rT}e^{-rT}-Ke^{-rT}\\ &= S_0-Ke^{-rT}\end{aligned}$$

This is the pricing formula for a forward contract. Such a derivative has no optionality.

CONCEPT REFRESHER 5.3: FORWARD CONTRACTS

A forward contract is an agreement between two parties to buy or sell an asset at a specific price K on a future date T. Once a forward contract is entered into, both parties have the obligation to satisfy the contract on its expiration date. Forward contracts are generally satisfied in one of two ways: delivery or cash settlement.

In the case of delivery, at expiration of the contract on date T, the seller delivers the asset underlying the forward contract to the purchaser at price K. In the case of cash settlement, at expiration of the contract on date T, if the price of the underlying asset S exceeds K, then the seller of the contract will make the payment in the amount of $S-K$ to the party who is long the contract. If $S<K$ at time T, the purchaser of the contract will make the payment $K-S$ to the contract seller. Accordingly, the payoff structure of a forward contract is linear, mimicking the payoff of an option without the kink and flattening (see Concept Refresher 5.1).

The symmetry of this exposure profile differentiates forward contracts from options, as only the seller of the option has a contractual obligation.

Forward contracts and futures contracts are similar but have slightly different terms. First, futures contracts are standardized and exchange-traded, whereas forward contracts are customizable by both parties. Futures contracts require daily margining, wherein parties post margin to each other depending on changes in the underlying asset price. This is not required in forward contracts, as margin requirements are part of the customization.

The expression for $E_N\left[\max(S_T - K, 0)\right]$ is similar. Since at expiration a call option only has value if the underlying stock price is greater than the strike price, we adjust the expression to focus only on values of S_T that are greater than K. We get

$$E_N\left[\max(S_T - K, 0)\right] = \int_K^{\infty} (s_T - K)\phi(s_T)\,ds_T \tag{96}$$

The difference between Equations 95 and 96 can be thought of as representing the optionality in a call option. Plugging Equation 96 into Equation 93, the value of a European call option can be written as

$$c_t = e^{-r(T-t)} \int_K^{\infty} (s_T - K)\phi(s_T)\,ds_T \tag{97}$$

Next we perform a change of variable. A change of variable is when variables from your original function are replaced with functions consisting of other variables. After the transformation, one hopes that the equation becomes easier. The change of variable we apply[15] requires us to define $S_T = e^{Zv+m}$. Rearranging, we see that $Z = \frac{\ln(S_T) - m}{v}$. We can think of the random variable Z as standardizing $\ln(S_T)$ where m and v are the mean and standard deviation of $\ln(S_T)$, respectively, and z is a realization of Z. Note that since S_T is the term of integration, the transformation also applies to the integration limits. This means the lower limit K becomes $\frac{\ln(K) - m}{v}$, while the upper limit is unchanged since $\lim_{S_T \to \infty} \frac{\ln(S_T) - m}{v} = \infty$. Equation 97 becomes

$$\begin{aligned} c_t &= e^{-r(T-t)} \int_{\frac{\ln(K)-m}{v}}^{\infty} \left(e^{(zv+m)} - K\right)\phi(z)\,dz \\ &= e^{-r(T-t)} \left[\int_{\frac{\ln(K)-m}{v}}^{\infty} e^{(zv+m)}\phi(z)\,dz - K \int_{\frac{\ln(K)-m}{v}}^{\infty} \phi(z)\,dz \right] \end{aligned} \tag{98}$$

We can collapse the second integral from Equation 98 to

$$K\int_{\frac{\ln(K)-m}{v}}^{\infty}\phi(z)dz = K\left[1-\int_{-\infty}^{\frac{\ln(K)-m}{v}}\phi(z)dz\right]$$

$$= K\left[1-\Phi\left(\frac{\ln(K)-m}{v}\right)\right]$$

Let's unpack this. Remember, $\phi(z)$ is a PDF, the integral of which is a CDF. The CDF $\Phi\left(\frac{\ln(K)-m}{v}\right)$ gives the probability that the random variable Z is less than or equal to $\frac{\ln(K)-m}{v}$. Thus, $1-\Phi\left(\frac{\ln(K)-m}{v}\right)$ gives the probability that Z is greater than $\frac{\ln(K)-m}{v}$. Because S_T is lognormally distributed, $Z=\frac{\ln(S_T)-m}{v}$ follows a standard normal distribution, so the CDF $\Phi(\cdot)$ here refers to that of the standard normal distribution.

What values should we use for mean m and standard deviation v when evaluating $\Phi\left(\frac{\ln(K)-m}{v}\right)$? In Equation 71 from Section 4.2.4, we established that

$$\ln(S_T)\sim N\left[\ln(S_t)+\left(\mu-\frac{\sigma^2}{2}\right)(T-t),\sigma^2(T-t)\right]$$

Plugging in these mean and variance values, we can write $\frac{\ln(K)-m}{v}$ as

$$\frac{\ln(K)-\left[\ln(S_t)+\left(\mu-\frac{\sigma^2}{2}\right)(T-t)\right]}{\sqrt{\sigma^2(T-t)}}$$

Recall that in a risk-neutral world, $\mu=r$. We can replace μ with r and simplify to get

$$\frac{\ln(K)-\left[\ln(S_t)+\left(r-\frac{\sigma^2}{2}\right)(T-t)\right]}{\sigma\sqrt{(T-t)}}=\frac{\ln(K)-\ln(S_t)-\left(r-\frac{\sigma^2}{2}\right)(T-t)}{\sigma\sqrt{(T-t)}}$$

Since the standard normal distribution is symmetric, it is true that $1-\Phi\left(\frac{\ln(K)-m}{d}\right)=\Phi\left(-\frac{\ln(K)-m}{d}\right)$. Multiplying through by -1:

$$\frac{-\ln(K)+\ln(S_t)+\left(r-\frac{\sigma^2}{2}\right)(T-t)}{\sigma\sqrt{(T-t)}}=\frac{\ln(S_t)-\ln(K)+\left(r-\frac{\sigma^2}{2}\right)(T-t)}{\sigma\sqrt{(T-t)}}$$

$$=\frac{\ln\left(\frac{S_t}{K}\right)+\left(r-\frac{\sigma^2}{2}\right)(T-t)}{\sigma\sqrt{(T-t)}}$$

Look familiar? This is just d_2 from Equation 91! Defining this term as d_2, we can rewrite the value of a European call option from Equation 98 as

$$\begin{aligned} c_t &= e^{-r(T-t)}\left[\int_{\frac{\ln(K)-m}{v}}^{\infty} e^{(zv+m)}\phi(z)dz - K\int_{\frac{\ln(K)-m}{v}}^{\infty}\phi(z)dz\right] \\ &= e^{-r(T-t)}\left[\int_{\frac{\ln(K)-m}{v}}^{\infty} e^{(zv+m)}\phi(z)dz - K\left[1-\Phi\left(\frac{\ln(K)-m}{v}\right)\right]\right] \qquad (99) \\ &= e^{-r(T-t)}\left[\int_{\frac{\ln(K)-m}{v}}^{\infty} e^{(zv+m)}\phi(z)dz - K\Phi(d_2)\right] \end{aligned}$$

Next we can move to the first integral term in Equation 99, $\int_{\frac{\ln(K)-m}{v}}^{\infty} e^{(zv+m)}\phi(z)dz$.

Let's start by combining and simplifying $e^{(zv+m)}\phi(z)$. Recall that $\phi(\cdot)$ is the PDF for the standard normal distribution. Following from Endnote 4 in Section 4 we have

$$e^{(zv+m)}\phi(z) = e^{(zv+m)}\frac{1}{\sqrt{2\pi}}\exp\left[-\frac{1}{2}z^2\right]$$

$$= \frac{1}{\sqrt{2\pi}}e^{zv+m-\frac{1}{2}z^2}$$

$$= \frac{1}{\sqrt{2\pi}}e^{\frac{2zv+2m-z^2}{2}}$$

$$= \frac{1}{\sqrt{2\pi}}e^{\frac{-(z-v)^2+2m+v^2}{2}}$$

since $-(z-v)^2 = -(z-v)(z-v) = -\left[z^2 - 2vz + v^2\right] = -z^2 + 2vz - v^2$. Then splitting the exponential into two terms gives us

$$= \frac{1}{\sqrt{2\pi}}e^{\frac{-(z-v)^2}{2}}e^{\frac{2m+v^2}{2}}$$

$$= e^{m+\frac{v^2}{2}}\left[\frac{1}{\sqrt{2\pi}}e^{\frac{-(z-v)^2}{2}}\right]$$

$$= e^{m+\frac{v^2}{2}}\left[\phi(z-v)\right] \tag{100}$$

Plugging Equation 100 into the formula for the price of a European call from Equation 99, we get

$$c_t = e^{-r(T-t)}\left[\int_{\frac{\ln(K)-m}{v}}^{\infty} e^{(zv+m)}\phi(z)dz - K\Phi(d_2)\right]$$

$$= e^{-r(T-t)}\left[\int_{\frac{\ln(K)-m}{v}}^{\infty} e^{m+\frac{v^2}{2}}\left[\phi(z-v)\right]dz - K\Phi(d_2)\right] \tag{101}$$

$$= e^{-r(T-t)}\left[e^{m+\frac{v^2}{2}}\int_{\frac{\ln(K)-m}{v}}^{\infty} \phi(z-v)dz - K\Phi(d_2)\right]$$

As before, we can write the integral from Equation 101 as

$$\int_{\frac{\ln(K)-m}{v}}^{\infty} \phi(z-v)dz \tag{102}$$

$$=1-\int_{-\infty}^{\frac{\ln(K)-m}{v}} \phi(z-v)dz$$

$$=1-\Phi\left(\frac{\ln(K)-m}{v}-v\right)$$

$$=\Phi\left(\frac{-\ln(K)+m}{v}+v\right) \tag{103}$$

Noting once again from Equation 71 in Section 4.2.4 that

$$\ln(S_T)\sim N\left[\ln(S_t)+\left(\mu-\frac{\sigma^2}{2}\right)(T-t),\sigma^2(T-t)\right] \tag{104}$$

and that in a risk-neutral world $\mu=r$, we can substitute the values of m and v from Equation 104 into Equation 103 to get

$$\Phi\left(\frac{-\ln(K)+m}{v}+v\right)$$

$$=\Phi\left(\frac{-\ln(K)+\ln(S_t)+\left(r-\frac{\sigma^2}{2}\right)(T-t)}{\sqrt{\sigma^2(T-t)}}+\sqrt{\sigma^2(T-t)}\right)$$

$$=\Phi\left(\frac{\ln\left(\frac{S_t}{K}\right)+\left(r-\frac{\sigma^2}{2}\right)(T-t)}{\sqrt{\sigma^2(T-t)}}+\sqrt{\sigma^2(T-t)}\right)$$

We note quickly that this is equivalent to $\Phi\left(d_2+\sqrt{\sigma^2(T-t)}\right)$, which is consistent with the definition of d_2 from Equation 91 that says $d_2=d_1-\sqrt{\sigma^2(T-t)}$. To see this, we continue to simplify

$$\Phi\left(\frac{\ln\left(\frac{S_t}{K}\right)+\left(r-\frac{\sigma^2}{2}\right)(T-t)}{\sqrt{\sigma^2(T-t)}}+\sqrt{\sigma^2(T-t)}\frac{\sqrt{\sigma^2(T-t)}}{\sqrt{\sigma^2(T-t)}}\right)$$

$$=\Phi\left(\frac{\ln\left(\frac{S_t}{K}\right)+\left(r-\frac{\sigma^2}{2}\right)(T-t)}{\sqrt{\sigma^2(T-t)}}+\frac{\sigma^2(T-t)}{\sqrt{\sigma^2(T-t)}}\right)$$

$$=\Phi\left(\frac{\ln\left(\frac{S_t}{K}\right)+\left(r-\frac{\sigma^2}{2}\right)(T-t)+\sigma^2(T-t)}{\sqrt{\sigma^2(T-t)}}\right)$$

$$=\Phi\left(\frac{\ln\left(\frac{S_t}{K}\right)+\left(r-\frac{\sigma^2}{2}+\sigma^2\right)(T-t)}{\sqrt{\sigma^2(T-t)}}\right)$$

$$=\Phi\left(\frac{\ln\left(\frac{S_t}{K}\right)+\left(r+\frac{\sigma^2}{2}\right)(T-t)}{\sqrt{\sigma^2(T-t)}}\right)$$

$$=\Phi(d_1)$$

We can now restate Equation 102 as

$$\int_{\frac{\ln(K)-m}{v}}^{\infty}\phi(z-v)dz=\Phi(d_1) \tag{105}$$

and use Equation 105 with the price of a European call option from Equation 101 to get

$$c_t = e^{-r(T-t)}\left[e^{m+\frac{v^2}{2}}\int_{\frac{\ln(K)-m}{v}}^{\infty}\phi(z-v)dz - K\Phi(d_2)\right]$$
$$= e^{-r(T-t)}\left[e^{m+\frac{v^2}{2}}\Phi(d_1) - K\Phi(d_2)\right] \tag{106}$$

Plugging in the values of m and v from Equation 104 into Equation 106 and remembering one last time that in a risk-neutral world $\mu = r$, we get

$$c_t = e^{-r(T-t)}\left[\exp\left(\ln(S_t)+\left(r-\frac{\sigma^2}{2}\right)(T-t)+\frac{\sigma^2(T-t)}{2}\right)\Phi(d_1) - K\Phi(d_2)\right]$$
$$= e^{-r(T-t)}\left[S_t\exp\left(r(T-t)-\frac{\sigma^2(T-t)}{2}+\frac{\sigma^2(T-t)}{2}\right)\Phi(d_1) - K\Phi(d_2)\right]$$
$$= e^{-r(T-t)}\left[S_t e^{r(T-t)}\Phi(d_1) - K\Phi(d_2)\right]$$

All that's left now is simple algebra. Distributing the term $e^{-r(T-t)}$, we get

$$e^{-r(T-t)}S_t e^{r(T-t)}\Phi(d_1) - e^{-r(T-t)}K\Phi(d_2)$$
$$= S_t\Phi(d_1) - K\Phi(d_2)e^{-r(T-t)}$$

And we're done! We have walked through every step required for deriving the BSM pricing formula under the risk-neutral assumption. Next, we'll explore an alternative way to get to the same place.

5.3.2 Binomial Tree

A binomial tree starts from a single node and branches to future nodes. It can be used to represent price movements of financial assets over discrete time increments. Figure 5.3 provides a basic example.

S_0 represents the price of some underlying security at time t_0 (today), which then branches off into two possible future values one time increment later. An up-move of proportional magnitude U brings the security price to the value S_u, while a down-move of proportional magnitude D brings the security to the value S_d.[16] The branching continues at each

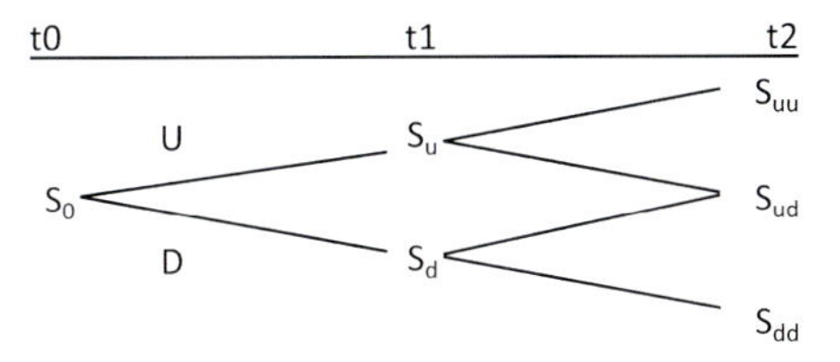

FIGURE 5.3 Binomial tree.

future node. The tree ends at time T, which in Figure 5.3 is time $T = 2$ (denoted t_2). There are, therefore, two time increments $(n = 2)$ between the beginning and end of the tree, each of length T/n.[17]

Recall that the expected value at expiration of a European call option on a non-dividend paying stock is

$$E\left[\max(S_T - K, 0)\right]$$

where again S_T is the price of the underlying stock at option expiration (time T) and K is the strike price. We can now describe S_T in terms of a binomial tree. To do so, replace S_T with $S_0 U^m D^{n-m}$. In other words, over n total time increments, the price of the underlying security will result from m up movements of proportional magnitude U and $n - m$ down movements of proportional magnitude D. As we are about to show, when the number of time increments between t_0 and T approaches infinity ($n \to \infty$, meaning $T/n \to 0$), one may recover the BSM model from the binomial tree.

Replacing S_T with $S_0 U^m D^{n-m}$, the expected value of a European call option becomes

$$E\left[\max\left(S_0 U^m D^{n-m} - K, 0\right)\right] \tag{107}$$

Now if you're wondering why it's called a "binomial" tree, it's because the probability of an up or down-move is ruled by the binomial distribution. So out of n total moves, the probability of m up-moves and $n - m$ down moves is

$$\frac{n!}{(n-m)!m!} p^m (1-p)^{n-m}$$

where p is the probability of an up-move and $(1 - p)$ is the probability of a down-move.

CONCEPT REFRESHER 5.4: BINOMIAL DISTRIBUTION

To better understand the binomial distribution, we derive it from scratch. To begin, we define a Bernoulli trial. A Bernoulli trial is a random experiment where the outcome is binary. For now, we will refer to these binary outcomes as success or failure. The probability of success (and by connection the probability of failure) is constant for every trial.

Since the probabilities of success and failure are constant, the probability of any particular sequence of events is trivial to calculate. For example, denote the probability of success p and the probability of failure $1-p$. The probability of a sequence of two consecutive successes is $p \times p = p^2$. Similarly, the probability of a success followed by a failure is $p \times (1-p)$. Generalizing, out of a total of n trials, one might be inclined to say the probability of exactly x successes is

$$p^x (1-p)^{n-x}$$

This is intuitive, since if there are x successes, there must be $n-x$ failures. Unfortunately, we must add some complexity. Say there are $n=4$ trials and we want to know the probability of exactly $x=2$ successes. There is more than one way to achieve two successes. For example, if S stands for success and F for failure, we might have

Sequence	Events
1	S,S,F,F
2	S,F,S,F
3	S,F,F,S
4	F,S,S,F
5	F,S,F,S
6	F,F,S,S

There are six different ways to get exactly two successes in four trials. To generalize this, we must have a universal way to figure out how many sequences achieve x successes in n total trials. The answer lies in basic combinatorics. Notice that each of sequences one through six achieves exactly two successes. The particular sequence to achieve two successes is irrelevant. Since order does not matter, we want to use a *combination*. To understand combinations, we start with the simpler *permutation*.

Permutations are used to understand how many ways members of a set can be arranged. Take a set of playing cards. There are 52 members in this set. Suppose we're interested in understanding how many different ways we could draw four cards. Let's start by assuming that the order in which the cards are drawn matters:

$$\underline{52} \quad \underline{51} \quad \underline{50} \quad \underline{49}$$

For the first draw, all 52 cards can be drawn. Once the first card is drawn, only 51 cards remain. On each successive draw, the number of available cards declines by one. The total number of permutations of four cards is then $52 \times 51 \times 51 \times 49$. Denoting the total number of cards $n = 52$ and the number of draws $x = 4$, we can restate the permutation calculation as

$$n(n-1)(n-2)(n-3)$$

$$= \frac{n(n-1)(n-2)(n-3)(n-4)(n-5)\ldots(n-(n-1))}{(n-4)(n-5)\ldots(n-(n-1))}$$

$$= \frac{n!}{(n-x)!}$$

Now what if we didn't care about the order in which the cards are drawn? Then some of the permutations would be repetitive. For example, the following two possible draws represent two different permutations but only one combination.

To find the total number of combinations, we must divide the total number of permutations by the number of hands that are different permutations but the same combination. To find the number of different permutations that represent the same combination, in this example, we need to ask how many different ways four cards can be arranged. This is just another permutation, so we know the answer is $4!$. Generally, x members can be arranged $x!$ number of ways. To account for these repetitive permutations, we simply divide by $x!$. The formula for the total number of combinations is then

$$\frac{n!}{(n-x)!} \frac{1}{x!} = \frac{n!}{(n-x)!x!}$$

In the case of Bernoulli trials, the outcome from the first trial does not affect the outcome from the second trial. This means order does not matter. Accordingly, we want to know the total number of combinations that result in a particular number of successes. Multiplying the probability of x successes by the number of combinations that lead to x successes, we arrive at

$$\frac{n!}{(n-x)!x!}p^x(1-p)^{n-x}$$

This is the binomial distribution! It provides the probability of exactly x successes out of n total Bernoulli trials. Let's now find the expected value and variance of a binomially distributed (discrete) random variable B with probability of success p. Taking the expectation gives us

$$E(B)=\sum_{b=0}^{n}bp(b)$$

$$=\sum_{b=0}^{n}b\frac{n!}{(n-b)!b!}p^b(1-p)^{n-b}$$

Since the whole term equals zero for $b=0$, we can start the summation at $b=1$:

$$E(B)=\sum_{b=1}^{n}b\frac{n!}{(n-b)!b!}p^b(1-p)^{n-b}$$

$$=\sum_{b=1}^{n}n\frac{(n-1)!}{(n-b)!(b-1)!}p^b(1-p)^{n-b}$$

$$=\sum_{b=1}^{n}np\frac{(n-1)!}{(n-b)!(b-1)!}p^{b-1}(1-p)^{n-b}$$

$$=np\sum_{b=1}^{n}\frac{(n-1)!}{(n-b)!(b-1)!}p^{b-1}(1-p)^{n-b}$$

Notice that the combination is now evaluating $x=b-1$ for a total of $n-1$ trials, since

$$\frac{(n-1)!}{\left[(n-1)-(b-1)\right]!(b-1)!}=\frac{(n-1)!}{(n-b)!(b-1)!}$$

and the probabilities reflect the probability of achieving $x=b-1$ in $n-1$ trials since

$$p^{b-1}(1-p)^{[(n-1)-(b-1)]}=p^{b-1}(1-p)^{n-b}$$

This means the summation adds to one, since all probabilities are included! So

$$E(B) = np\sum_{b=1}^{n}\frac{(n-1)!}{(n-b)!(b-1)!}p^{b-1}(1-p)^{n-b} = np$$

Now for the variance, we have

$$Var(B) = E\left(B^2\right) - E(B)^2$$

The latter term is simply $E(B)^2 = (np)^2 = n^2p^2$. The first term requires a bit more work. Written out, and following a similar path as the expected value, we have (starting again with $b=1$ in the summation since $b=0$ reduces the entire summand to zero):

$$\begin{aligned}
E\left(B^2\right) &= \sum_{b=1}^{n} b^2 p(b) \\
&= \sum_{b=1}^{n} b^2 \frac{n!}{(n-b)!b!}p^b(1-p)^{n-b} \\
&= \sum_{b=1}^{n} b \frac{n!}{(n-b)!(b-1)!}p^b(1-p)^{n-b} \\
&= np\sum_{b=1}^{n} b \frac{(n-1)!}{(n-b)!(b-1)!}p^{b-1}(1-p)^{n-b} \\
&= np\sum_{b=1}^{n} (b-1+1) \frac{(n-1)!}{(n-b)!(b-1)!}p^{b-1}(1-p)^{n-b} \\
&= np\sum_{b=1}^{n} (b-1) \frac{(n-1)!}{(n-b)!(b-1)!}p^{b-1}(1-p)^{n-b} \\
&\quad + np\sum_{b=1}^{n} \frac{(n-1)!}{(n-b)!(b-1)!}p^{b-1}(1-p)^{n-b} \\
&= np\left[(n-1)p\right] + np(1) \\
&= np\left[pn - p\right] + np \\
&= n^2p^2 - np^2 + np
\end{aligned}$$

Putting it all together gives us

$$Var(B) = E(B^2) - E(B)^2 = n^2p^2 - np^2 + np - n^2p^2$$
$$= np - np^2$$
$$= np(1-p)$$

Summarizing, the expected value and variance of a binomial distributed random variable are $E(B) = np$ and $Var(B) = np(1-p)$, respectively.

Now let's spend a moment solving for p, as it will help us in a bit. Consider again Equation 70 from Section 4.2.4:

$$S_t = S_0 \exp\left[\left(\mu - \frac{\sigma^2}{2}\right)\Delta t + \sigma\varepsilon_t\sqrt{\Delta t}\right]$$

We know that S_t is lognormally distributed. The expected value (mean) of a lognormally distributed random variable $X \sim LN(\mu, \sigma^2)$ is

$$E(X) = \exp\left[\mu + \frac{\sigma^2}{2}\right]$$

The expected value of S_t is then

$$E(S_t) = E\left(S_0 \exp\left[\left(\mu - \frac{\sigma^2}{2} + \frac{\sigma^2}{2}\right)\Delta t + \sigma\varepsilon_t\sqrt{\Delta t}\right]\right) \tag{108}$$
$$= S_0 e^{\mu\Delta t}$$

since $E\left(\sigma\varepsilon_t\sqrt{\Delta t}\right) = \sigma E(\varepsilon_t)\sqrt{\Delta t} = 0$.

Writing the expected value of S_t in terms of the binomial tree, we have

$$E(S_t) = S_0\left[pU + (1-p)D\right] \tag{109}$$

Setting Equation 108 equal to Equation 109 gives us

$$S_0 e^{\mu\Delta t} = S_0\left[pU + (1-p)D\right]$$
$$e^{\mu\Delta t} = pU + D - pD$$
$$e^{\mu\Delta t} - D = pU - pD$$

$$e^{\mu\Delta t} - D = p[U - D]$$

$$p = \frac{e^{\mu\Delta t} - D}{U - D} \tag{110}$$

We now define the up and down factors U and D, respectively, to match volatility.[18] Recall from Equation 59 in Section 4.2.2 that the variance over time increment Δt of a stock price following a generalized Wiener process is $\sigma^2 \Delta t$. In terms of the binomial tree, the expected return of S_t is

$$E(S_t) = p(U-1) + (1-p)(D-1)$$

since $U-1$ is the return from an up-move and $D-1$ is the return from a down-move (see Endnote 16). The variance of S_t in terms of the binomial tree is then

$$\begin{aligned} Var(S_t) &= E(S_t^2) - E(S_t)^2 \\ &= p(U-1)^2 + (1-p)(D-1)^2 - \left[p(U-1) + (1-p)(D-1)\right]^2 \end{aligned} \tag{111}$$

Setting Equation 111 equal to the variance from the generalized Wiener process gives us

$$p(U-1)^2 + (1-p)(D-1)^2 - \left[p(U-1) + (1-p)(D-1)\right]^2 = \sigma^2 \Delta t \tag{112}$$

Substituting Equation 110 into Equation 112 gives us

$$\begin{aligned} &\frac{e^{\mu\Delta t} - D}{U - D}(U-1)^2 + \left(1 - \frac{e^{\mu\Delta t} - D}{U - D}\right)(D-1)^2 \\ &- \left[\frac{e^{\mu\Delta t} - D}{U - D}(U-1) + \left(1 - \frac{e^{\mu\Delta t} - D}{U - D}\right)(D-1)\right]^2 = \sigma^2 \Delta t \end{aligned} \tag{113}$$

Sparing the reader pages of algebra, the solution to Equation 113 for the up and down factors U and D, respectively, is

$$U = e^{\sigma\sqrt{\Delta t}} \tag{114}$$

and

$$D = \frac{1}{U} = e^{-\sigma\sqrt{\Delta t}} \tag{115}$$

These values of U and D equate to spacing the discrete nodes on the tree according to the volatility of the log price. Such an approach ensures that the tree recombines (i.e., that $S_{ud} = S_{du}$) and that values can be derived formulaically rather than requiring simulation. Substituting Equations 114 and 115 into Equation 110, we can rewrite the probability of an up-move as

$$\begin{aligned} p &= \frac{e^{\mu \Delta t} - D}{U - D} \\ &= \frac{e^{\mu \Delta t} - e^{-\sigma\sqrt{\Delta t}}}{e^{\sigma\sqrt{\Delta t}} - e^{-\sigma\sqrt{\Delta t}}} \end{aligned} \tag{116}$$

Since we know that in the risk-neutral world the drift term μ is equal to the risk-free rate r and that $\Delta t = T/n$ on the binomial tree, the risk-neutral probability of an up-move is then

$$p = \frac{e^{rT/n} - e^{-\sigma\sqrt{T/n}}}{e^{\sigma\sqrt{T/n}} - e^{-\sigma\sqrt{T/n}}} \tag{117}$$

Recall that the expected value of a discrete random variable X is written as

$$E(X) = \sum_{i=1}^{n} x_i f(x_i)$$

This is just a weighted average of the different possible values of X, with $f(x_i)$ being the probability (the weight) of the ith observation of x. Applying this to the expected value of a European call option, $\max\left(S_0 U^m D^{n-m} - K, 0\right)$ takes the place of X and $\frac{n!}{(n-m)!m!} p^m (1-p)^{n-m}$ takes the place of $f(x)$. The expected value of the European call option from Equation 107 becomes

$$E\left[\max\left(S_0 U^m D^{n-m} - K, 0\right)\right] = \sum_{m=0}^{n} \max\left(S_0 U^m D^{n-m} - K, 0\right) \frac{n!}{(n-m)!m!} p^m (1-p)^{n-m}$$

This is the expected value at time T, which must then be discounted back to time 0. Since the binomial tree is still assumed to operate within a risk-neutral world, it is appropriate to discount at the risk-free rate r. The value of a European call option at time t_0 is then

$$c_{t_0} = e^{-rT} \sum_{m=0}^{n} \max\left(S_0 U^m D^{n-m} - K, 0\right) \frac{n!}{(n-m)!m!} p^m (1-p)^{n-m}$$

Because of the max function, all outcomes in which $S_0U^mD^{n-m} < K$ are zero. This represents the call option expiring worthless. The value of the call option must only stem from scenarios in which $S_0U^mD^{n-m} > K$, reflecting the condition that the call option expires in-the-money. We can rewrite this condition to solve for m, that is, to figure out exactly how many up-moves we need to ensure $S_0U^mD^{n-m} > K$:

$$S_0U^mD^{n-m} > K$$

$$\frac{S_0}{K}U^mD^{n-m} > 1$$

$$\ln\left(\frac{S_0}{K}U^mD^{n-m}\right) > \ln(1)$$

$$\ln\left(\frac{S_0}{K}\right) + \ln\left(U^m\right) + \ln\left(D^{n-m}\right) > 0$$

$$\ln\left(\frac{S_0}{K}\right) + m\ln(U) + (n-m)\ln(D) > 0$$

$$\ln\left(\frac{S_0}{K}\right) > -m\ln(U) - (n-m)\ln(D) \tag{118}$$

As defined in Equations 114 and 115, $U = e^{\sigma\sqrt{T/n}}$ and $D = e^{-\sigma\sqrt{T/n}}$. Therefore, we can substitute these values for U and D into the condition from Equation 118 to get

$$\ln\left(\frac{S_0}{K}\right) > -m\ln\left(e^{\sigma\sqrt{T/n}}\right) - (n-m)\ln\left(e^{-\sigma\sqrt{T/n}}\right)$$

$$\ln\left(\frac{S_0}{K}\right) > -m\sigma\sqrt{T/n} - (n-m)\left(-\sigma\sqrt{T/n}\right)$$

$$\ln\left(\frac{S_0}{K}\right) > -m\sigma\sqrt{T/n} + (n-m)\left(\sigma\sqrt{T/n}\right)$$

$$\ln\left(\frac{S_0}{K}\right) > -m\sigma\sqrt{T/n} + n\sigma\sqrt{T/n} - m\sigma\sqrt{T/n}$$

$$\ln\left(\frac{S_0}{K}\right) > n\sigma\sqrt{T/n} - 2m\sigma\sqrt{T/n}$$

$$\frac{\ln\left(\frac{S_0}{K}\right)}{\sigma\sqrt{T/n}} > n - 2m$$

$$\frac{\ln\left(\frac{S_0}{K}\right)}{2\sigma\sqrt{T/n}} > \frac{n}{2} - m$$

$$m > \frac{n}{2} - \frac{\ln\left(\frac{S_0}{K}\right)}{2\sigma\sqrt{T/n}}$$

We have now restated the condition in terms of the number of up-moves m. There must be more than $\frac{n}{2} - \frac{\ln(S_0/K)}{2\sigma\sqrt{T/n}}$ up-moves for the call option to have a positive value at expiration. Denoting the threshold α, we have

$$\alpha = \frac{n}{2} - \frac{\ln\left(\frac{S_0}{K}\right)}{2\sigma\sqrt{T/n}} \tag{119}$$

We can now rewrite the value of the European call option as

$$\begin{aligned} c_{t_0} &= e^{-rT}\sum_{m=0}^{n}\max\left(S_0U^mD^{n-m} - K, 0\right)\frac{n!}{(n-m)!m!}p^m(1-p)^{n-m} \\ &= e^{-rT}\sum_{m>\alpha}\left(S_0U^mD^{n-m} - K\right)\frac{n!}{(n-m)!m!}p^m(1-p)^{n-m} \end{aligned} \tag{120}$$

Notice that the set over which we take the summation is narrowed from $m = [0, n]$ to $m > \alpha$. This change allows us to drop the max function, since we know that $S_0U^mD^{n-m} - K > 0$ for all $m > \alpha$. Now let's focus on the red term in Equation 120. This term gives the probability of exactly m up-moves. And since we've restricted the summation region to $m > \alpha$, we're only calculating the probabilities for m's that are high enough that the option has value at option expiration (i.e., m's high enough that the blue term in Equation 120 is greater than zero). Distributing the red term to the blue term, we get

$$\begin{aligned} c_{t_0} &= e^{-rT}\sum_{m>\alpha}\left[S_0U^mD^{n-m}\frac{n!}{(n-m)!m!}p^m(1-p)^{n-m} - K\frac{n!}{(n-m)!m!}p^m(1-p)^{n-m}\right] \\ &= e^{-rT}\left[\sum_{m>\alpha}S_0U^mD^{n-m}\frac{n!}{(n-m)!m!}p^m(1-p)^{n-m} - \sum_{m>\alpha}K\frac{n!}{(n-m)!m!}p^m(1-p)^{n-m}\right] \\ &= e^{-rT}\left[S_0\sum_{m>\alpha}U^mD^{n-m}\frac{n!}{(n-m)!m!}p^m(1-p)^{n-m} - K\sum_{m>\alpha}\frac{n!}{(n-m)!m!}p^m(1-p)^{n-m}\right] \end{aligned} \tag{121}$$

To make this easier on the eyes, we can rewrite the green and purple terms in Equation 121, respectively, as

$$Q_1 = \sum_{m>\alpha} U^m D^{n-m} \frac{n!}{(n-m)!m!} p^m (1-p)^{n-m}$$

$$Q_2 = \sum_{m>\alpha} \frac{n!}{(n-m)!m!} p^m (1-p)^{n-m}$$

This allows us to rewrite Equation 121 as

$$c_{t_0} = e^{-rT} \left(S_0 Q_1 - K Q_2 \right) \tag{122}$$

Q_2 has a very convenient interpretation. It is the probability that $m > \alpha$, meaning it's also the probability that $S_0 U^m D^{n-m} > K$, equivalently that $S_T > K$. In other words, Q_2 is the probability that the European call option will be in-the-money at expiration.

Remember, with n total time increments and probability of an up-move p, the number of up-moves m is binomially distributed with mean np and standard deviation $\sqrt{np(1-p)}$ (see Concept Refresher 5.4). A convenient property of the binomial distribution is that it converges to the normal distribution asymptotically (as $n \to \infty$). So if we assume that n is sufficiently large, then we can approximate Q_2 with the normal distribution. That is,

$$\begin{aligned} Q_2 &= \sum_{m>\alpha} \frac{n!}{(n-m)!m!} p^m (1-p)^{n-m} \\ &\approx \Phi\left(\frac{np-\alpha}{\sqrt{np(1-p)}} \right) \end{aligned} \tag{123}$$

Again, Q_2 is the probability that $m > \alpha$. The term inside the parentheses in Equation 123 is nothing more than the standardized[19] distance of m from the threshold α. As in the prior section, the function $\Phi(\cdot)$ returns the probability that a standard normal random variable is less than $\frac{np-\alpha}{\sqrt{np(1-p)}}$, or equivalently the probability that $np > \alpha$. We derived in Equation 119 that $\alpha = \frac{n}{2} - \frac{\ln\left(\frac{S_0}{K}\right)}{2\sigma\sqrt{T/n}}$. Plugging this value of α into Equation 123 gives us

$$
\begin{aligned}
Q_2 &= \Phi\left(\frac{np-\alpha}{\sqrt{np(1-p)}}\right) \\
&= \Phi\left(\frac{np-\frac{n}{2}+\frac{\ln\left(\frac{S_0}{K}\right)}{2\sigma\sqrt{T/n}}}{\sqrt{np(1-p)}}\right) \\
&= \Phi\left(\frac{\frac{\ln\left(\frac{S_0}{K}\right)}{2\sigma\sqrt{T/n}}}{\sqrt{np(1-p)}}+\frac{np-\frac{n}{2}}{\sqrt{np(1-p)}}\right) \\
&= \Phi\left(\frac{\ln\left(\frac{S_0}{K}\right)}{2\sigma\sqrt{T/n}\sqrt{np(1-p)}}+\frac{n\left(p-\frac{1}{2}\right)}{\sqrt{np(1-p)}}\right) \\
&= \Phi\left(\frac{\ln\left(\frac{S_0}{K}\right)}{2\sigma\sqrt{Tp(1-p)}}+\frac{\sqrt{n}\left(p-\frac{1}{2}\right)}{\sqrt{p(1-p)}}\right)
\end{aligned}
\tag{124}
$$

Recall from Equation 117 that

$$
p=\frac{e^{rT/n}-e^{-\sigma\sqrt{T/n}}}{e^{\sigma\sqrt{T/n}}-e^{-\sigma\sqrt{T/n}}}
$$

The next step would be to substitute this equation for p into Equation 124 to solve for Q_2, but it's clear that the equation would become somewhat unwieldy. Instead, let's see what we can learn about how p behaves asymptotically (i.e., as $n \to \infty$). Taking the limit of p as n approaches infinity represents the collapsing of discrete time into continuous time. In other words, in the asymptote, the binomial tree has an infinite number of steps between times t_0 and T, each of infinitesimally small length.

$$\lim_{n\to\infty} p = \lim_{n\to\infty} \frac{e^{rT/n} - e^{-\sigma\sqrt{T/n}}}{e^{\sigma\sqrt{T/n}} - e^{-\sigma\sqrt{T/n}}}$$

$$= \lim_{n\to\infty} \frac{e^{-\sigma\sqrt{T/n}}}{e^{-\sigma\sqrt{T/n}}} \left(\frac{e^{rT/n+\sigma\sqrt{T/n}} - 1}{e^{2\sigma\sqrt{T/n}} - 1} \right)$$

$$= \lim_{n\to\infty} \left(\frac{e^{rT/n+\sigma\sqrt{T/n}} - 1}{e^{2\sigma\sqrt{T/n}} - 1} \right)$$

Using L'Hopital's rule,[20] we write

$$\lim_{n\to\infty} \left(\frac{e^{rT/n+\sigma\sqrt{T/n}} - 1}{e^{2\sigma\sqrt{T/n}} - 1} \right) = \lim_{n\to\infty} \left(\frac{\frac{d}{dn} e^{rT/n+\sigma\sqrt{T/n}} - 1}{\frac{d}{dn} e^{2\sigma\sqrt{T/n}} - 1} \right) \tag{125}$$

Solving the right-hand-side of Equation 125, we have

$$\lim_{n\to\infty} \left(\frac{\frac{d}{dn} e^{rT/n+\sigma\sqrt{T/n}} - 1}{\frac{d}{dn} e^{2\sigma\sqrt{T/n}} - 1} \right) = \lim_{n\to\infty} \left(\frac{\frac{d}{dn} e^{rTn^{-1}+\sigma\sqrt{T}n^{-\frac{1}{2}}} - 1}{\frac{d}{dn} e^{2\sigma\sqrt{T}n^{-\frac{1}{2}}} - 1} \right)$$

$$= \lim_{n\to\infty} \left(\frac{\left[e^{rT/n+\sigma\sqrt{T/n}} \right] \left[-\frac{rT}{n^2} - \frac{\sigma\sqrt{T}}{2n^{3/2}} \right]}{\left[e^{2\sigma\sqrt{T/n}} \right] \left[-\frac{\sigma\sqrt{T}}{n^{3/2}} \right]} \right)$$

$$= \lim_{n\to\infty} \left(\frac{e^{rT/n} e^{\sigma\sqrt{T/n}} \left[-\frac{rT}{n^2} - \frac{\sigma\sqrt{T}}{2n^{3/2}} \right]}{e^{\sigma\sqrt{T/n}} e^{\sigma\sqrt{T/n}} \left[-\frac{\sigma\sqrt{T}}{n^{3/2}} \right]} \right)$$

$$= \lim_{n\to\infty} \left(\frac{e^{rT/n} \left[-\frac{2rT - \sigma\sqrt{T}n^{1/2}}{2n^2} \right]}{e^{\sigma\sqrt{T/n}} \left[-\frac{\sigma\sqrt{T}}{n^{3/2}} \right]} \right)$$

$$= \lim_{n\to\infty}\left(e^{rT/n-\sigma\sqrt{T/n}} \frac{-\dfrac{2rT-\sigma\sqrt{T}n^{1/2}}{2n^2}}{-\dfrac{\sigma\sqrt{T}}{n^{3/2}}} \right)$$

$$= \lim_{n\to\infty}\left(\frac{-n^{3/2}\left[2rT+\sigma\sqrt{T}n^{1/2}\right]}{-2n^2\sigma\sqrt{T}} \right)$$

$$= \lim_{n\to\infty}\left(\frac{-2rT-\sigma\sqrt{T}n^{1/2}}{-2n^{1/2}\sigma\sqrt{T}} \right)$$

$$= \frac{-\sigma\sqrt{T}n^{1/2}}{-2n^{1/2}\sigma\sqrt{T}}$$

$$= \frac{1}{2}$$

since $\lim_{n\to\infty}\left(e^{rT/n-\sigma\sqrt{T/n}}\right)=e^0=1$. Thus, we can say

$$\lim_{n\to\infty} p = \frac{1}{2}$$

$$\lim_{n\to\infty} p(1-p) = \frac{1}{4} \tag{126}$$

Following a similar procedure, the limit for the term $\sqrt{n}\left(p-\frac{1}{2}\right)$ is

$$\lim_{n\to\infty}\sqrt{n}\left(p-\frac{1}{2}\right)=\frac{\left(r-\dfrac{\sigma^2}{2}\right)\sqrt{T}}{2\sigma} \tag{127}$$

Plugging Equations 126 and 127 into Equation 124 to solve for Q_2, we get

$$Q_2 = \Phi\left(\frac{\ln\left(\dfrac{S_0}{K}\right)}{2\sigma\sqrt{Tp(1-p)}} + \frac{\sqrt{n}\left(p-\dfrac{1}{2}\right)}{\sqrt{p(1-p)}} \right)$$

$$= \Phi\left(\frac{\ln\left(\frac{S_0}{K}\right)}{2\sigma\sqrt{T/4}} + \frac{\frac{\left(r - \frac{\sigma^2}{2}\right)\sqrt{T}}{2\sigma}}{\sqrt{1/4}}\right)$$

$$= \Phi\left(\frac{\ln\left(\frac{S_0}{K}\right)}{2\sigma\sqrt{T}\sqrt{1/4}} + \frac{\left(r - \frac{\sigma^2}{2}\right)\sqrt{T}}{\frac{2\sigma}{2}}\right)$$

$$= \Phi\left(\frac{\ln\left(\frac{S_0}{K}\right)}{\frac{2\sigma\sqrt{T}}{2}} + \frac{\left(r - \frac{\sigma^2}{2}\right)\sqrt{T}}{\sigma}\right)$$

$$= \Phi\left(\frac{\ln\left(\frac{S_0}{K}\right)}{\sigma\sqrt{T}} + \frac{\left(r - \frac{\sigma^2}{2}\right)T}{\sigma\sqrt{T}}\right)$$

$$= \Phi\left(\frac{\ln\left(\frac{S_0}{K}\right) + \left(r - \frac{\sigma^2}{2}\right)T}{\sigma\sqrt{T}}\right)$$

This should look familiar, as it is equivalent to $\Phi(d_2)$ from Equation 91. This proves that for the value of c_{t_0} from Equation 122, we can replace Q_2 with $\Phi(d_2)$. Stopping here, we can now say

$$c_{t_0} = e^{-rT}\left(S_0 Q_1 - K\Phi(d_2)\right) \tag{128}$$

Next, we move on to Q_1. Rewriting Q_1 (the green term from Equation 121), we have

$$\begin{aligned} Q_1 &= \sum_{m>\alpha} U^m D^{n-m} \frac{n!}{(n-m)!m!} p^m (1-p)^{n-m} \\ &= \sum_{m>\alpha} \frac{n!}{(n-m)!m!} (pU)^m \left[(1-p)D\right]^{n-m} \end{aligned} \tag{129}$$

Notice we have a new probability for moves up and down. Instead of $p^m(1-p)^{n-m}$ as it was with Q_2, we have $(pU)^m\left[(1-p)D\right]^{n-m}$. This new probability can be defined as

$$p^* = \frac{pU}{pU + (1-p)D} \tag{130}$$

meaning that

$$1 - p^* = \frac{(1-p)D}{pU + (1-p)D}$$

Rearranging these, we have

$$\begin{aligned} pU &= p^*\left[pU + (1-p)D\right] \\ (1-p)D &= (1-p^*)\left[pU + (1-p)D\right] \end{aligned}$$

We can now rewrite Q_1 from Equation 129 as

$$\begin{aligned} Q_1 &= \sum_{m>\alpha} \frac{n!}{(n-m)!m!} (pU)^m \left[(1-p)D\right]^{n-m} \\ &= \sum_{m>\alpha} \frac{n!}{(n-m)!m!} \left(p^*\left[pU + (1-p)D\right]\right)^m \left[(1-p^*)\left[pU + (1-p)D\right]\right]^{n-m} \\ &= \left[pU + (1-p)D\right]^m \left[pU + (1-p)D\right]^{n-m} \sum_{m>\alpha} \frac{n!}{(n-m)!m!} (p^*)^m \left[(1-p^*)\right]^{n-m} \\ &= \left[pU + (1-p)D\right]^{m+n-m} \sum_{m>\alpha} \frac{n!}{(n-m)!m!} (p^*)^m \left[(1-p^*)\right]^{n-m} \\ &= \left[pU + (1-p)D\right]^n \sum_{m>\alpha} \frac{n!}{(n-m)!m!} (p^*)^m \left[1-p^*\right]^{n-m} \end{aligned} \tag{131}$$

Notice that the pink term is the expected rate of return of the underlying stock over n time increments. Because we're in a risk-neutral world, we know that the expected rate of return on the underlying stock must tie back to the risk-free rate. Therefore,

$$pU + (1-p)D = e^{rT/n} \tag{132}$$

With this in mind, we write Equation 131 as

$$\begin{aligned} Q_1 &= \left[pU + (1-p)D\right]^n \sum_{m>\alpha} \frac{n!}{(n-m)!m!} (p^*)^m \left[1-p^*\right]^{n-m} \\ &= \left[e^{rT/n}\right]^n \sum_{m>\alpha} \frac{n!}{(n-m)!m!} (p^*)^m \left[1-p^*\right]^{n-m} \\ &= e^{rT} \sum_{m>\alpha} \frac{n!}{(n-m)!m!} (p^*)^m \left[1-p^*\right]^{n-m} \end{aligned} \tag{133}$$

We now have essentially the exact same form as we did for Q_2, except for the use of p^* instead of p and the e^{rT} term in front. We can apply the same normal approximation for Equation 133 as we did with Equation 123. That is,

$$\begin{aligned} Q_1 &= e^{rT} \sum_{m>\alpha} \frac{n!}{(n-m)!m!} (p^*)^m \left[1-p^*\right]^{n-m} \\ &\approx e^{rT} \Phi\left(\frac{np^* - \alpha}{\sqrt{np^*(1-p^*)}} \right) \\ &= e^{rT} \Phi\left(\frac{np^* - \frac{n}{2} + \frac{\ln\left(\frac{S_0}{K}\right)}{2\sigma\sqrt{T/n}}}{\sqrt{np^*(1-p^*)}} \right) \end{aligned}$$

$$=e^{rT}\Phi\left(\frac{\frac{\ln\left(\frac{S_0}{K}\right)}{2\sigma\sqrt{T/n}}}{\sqrt{np^*\left(1-p^*\right)}}+\frac{np^*-\frac{n}{2}}{\sqrt{np^*\left(1-p^*\right)}}\right)$$

$$=e^{rT}\Phi\left(\frac{\ln\left(\frac{S_0}{K}\right)}{2\sigma\sqrt{T/n}\sqrt{np^*\left(1-p^*\right)}}+\frac{n\left(p^*-\frac{1}{2}\right)}{\sqrt{np^*\left(1-p^*\right)}}\right)$$

$$=e^{rT}\Phi\left(\frac{\ln\left(\frac{S_0}{K}\right)}{2\sigma\sqrt{Tnp^*\left(1-p^*\right)/n}}+\frac{\sqrt{n}\left(p^*-\frac{1}{2}\right)}{\sqrt{p^*\left(1-p^*\right)}}\right)$$

$$=e^{rT}\Phi\left(\frac{\ln\left(\frac{S_0}{K}\right)}{2\sigma\sqrt{Tp^*\left(1-p^*\right)}}+\frac{\sqrt{n}\left(p^*-\frac{1}{2}\right)}{\sqrt{p^*\left(1-p^*\right)}}\right) \tag{134}$$

Restating p^* from Equation 130 while replacing U in the numerator with Equation 114, p in the numerator with Equation 117 and the denominator with Equation 132, we get

$$p^*=\frac{pU}{pU+\left(1-p\right)D}$$

$$=\left(\frac{e^{rT/n}-e^{-\sigma\sqrt{T/n}}}{e^{\sigma\sqrt{T/n}}-e^{-\sigma\sqrt{T/n}}}\right)\left(\frac{e^{\sigma\sqrt{T/n}}}{e^{rT/n}}\right)$$

Taking the same approach as we did to arrive at Equations 126 and 127 shows that

$$\lim_{n\to\infty}p^*\left(1-p^*\right)=\frac{1}{4} \tag{135}$$

$$\lim_{n\to\infty}\sqrt{n}\left(p^*-\frac{1}{2}\right)=\frac{\left(r+\frac{\sigma^2}{2}\right)\sqrt{T}}{2\sigma} \tag{136}$$

Substituting Equations 135 and 136 into Equation 134 gives us

$$Q_1 = e^{rT}\Phi\left(\frac{\ln\left(\frac{S_0}{K}\right)}{2\sigma\sqrt{Tp^*\left(1-p^*\right)}} + \frac{\sqrt{n}\left(p^* - \frac{1}{2}\right)}{\sqrt{p^*\left(1-p^*\right)}}\right)$$

$$= e^{rT}\Phi\left(\frac{\ln\left(\frac{S_0}{K}\right)}{2\sigma\sqrt{T/4}} + \frac{\frac{\left(r+\frac{\sigma^2}{2}\right)\sqrt{T}}{2\sigma}}{\sqrt{1/4}}\right)$$

$$= e^{rT}\Phi\left(\frac{\ln\left(\frac{S_0}{K}\right)}{2\sigma\sqrt{T}\sqrt{1/4}} + \frac{\left(r+\frac{\sigma^2}{2}\right)\sqrt{T}}{2\sigma/2}\right)$$

$$= e^{rT}\Phi\left(\frac{\ln\left(\frac{S_0}{K}\right)}{2\sigma\sqrt{T}/2} + \frac{\left(r+\frac{\sigma^2}{2}\right)T}{\sigma\sqrt{T}}\right)$$

$$= e^{rT}\Phi\left(\frac{\ln\left(\frac{S_0}{K}\right)}{\sigma\sqrt{T}} + \frac{\left(r+\frac{\sigma^2}{2}\right)T}{\sigma\sqrt{T}}\right)$$

$$= e^{rT}\Phi\left(\frac{\ln\left(\frac{S_0}{K}\right)+\left(r+\frac{\sigma^2}{2}\right)T}{\sigma\sqrt{T}}\right)$$

This should also look familiar, as the term in brackets is equivalent to d_1 from Equation 90. This proves that we can replace Q_1 with $e^{rT}\Phi(d_1)$ in the formula for c_{t_0} from Equation 128. We can now say

$$
\begin{aligned}
c_{t_0} &= e^{-rT}\left(S_0 Q_1 - K\Phi(d_2)\right) \\
&= e^{-rT}\left(S_0 e^{rT}\Phi(d_1) - K\Phi(d_2)\right) \\
&= S_0\Phi(d_1) - Ke^{-rT}\Phi(d_2)
\end{aligned}
$$

And we're done. We have proven the BSM formula from the binomial tree.

NOTES

1 This section is an adaptation of Cornuejols et al. (2018).

2 Notice that the first inequality implies that the amount invested in some bonds must be negative, meaning a short position in at least one bond. Conversely, the second inequality implies that the amount invested in some bonds must be positive, meaning a long position in at least one bond.

3 Taking expectation with respect to a probability measure can be thought of as taking a weighted average where the weights are probabilities from a particular distribution.

4 This discussion is instructive regarding what it means to operate in the risk-neutral world, where assets are expected to grow at the risk-free rate.

5 This section is an adaptation of multiple sources, including the original sources Black and Scholes (1973) and Merton (1974), as well as summaries from Hull (2018), Gupta (2014), and Tsay (2010).

6 There is evidence that stock prices cannot be modeled using GBM (e.g., see Lo and MacKinlay (1987)). For purposes of deriving the Black-Scholes-Merton differential equation and pricing model, we must assume stock prices *can* be modeled using GBM.

7 Since Ito's lemma is used to describe the process for the derivative G, its assumptions are inherited here as well. In particular, those from Endnote 11 in Section 4.

8 A *short* position in a security simply means borrowing the security and then selling it. You receive the price of the security now, but will eventually have to repurchase the security and return it to the lender. You would do this if you think the price of the security will decline in the future.

9 We also know that Ito's Lemma, as defined in Equation 66, has terms that are only deterministic in the limit (e.g., from Equation 64).

10 Arbitrage refers to earning profit in excess of the risk-free rate without taking any risk. Every risk-free security should provide a risk-free return, but nothing more. If a risk-free security provided return exceeding the risk-free rate, then market participants would purchase or sell securities to capture this increased return. In doing so, they would impact the prices of the securities in such a way as to eliminate the arbitrage opportunity.

11 For the risk-free portfolio to evolve at the risk-free rate, we must assume that that unlimited borrowing and lending at the risk-free rate is possible so as to eliminate all arbitrate opportunities. Any mispricings that result

in arbitrage opportunities incentivize market participants to act on them quickly. If borrowing rates are sufficiently higher than the risk-free rate, it may not be profitable to exploit arbitrage opportunities, potentially allowing mispricing to persist. If this were the case, it could no longer be guaranteed that the dynamically hedged portfolio will receive the risk-free return.

12 Rho and Gamma are more Greek terms. Rho is the change in the price of a derivative for a change in time. Gamma is the change in delta for a change in the price of the underlying security.

13 This section is an adaptation of multiple sources, including the original sources Black and Scholes (1973) and Merton (1974), as well as summaries from Hull (2018), Gupta (2014), and Tsay (2010).

14 Closed-form means an equation can be solved using standard operations (i.e., it does not require simulation). One can simply input values for each of the variables and solve.

15 It is also possible to apply a change of variable using $S_T = e^x$ and get to the same place, but this will not be shown. See Tsay (2010).

16 We say "proportional" because U and D will typically be represented as a percentage of the prior price. In other words, $S_u = S_0U$ and $S_d = S_0D$. If $S_0 = 50$, $U = 1.3$, $D = 0.7$, then the first part of tree will have the following form:

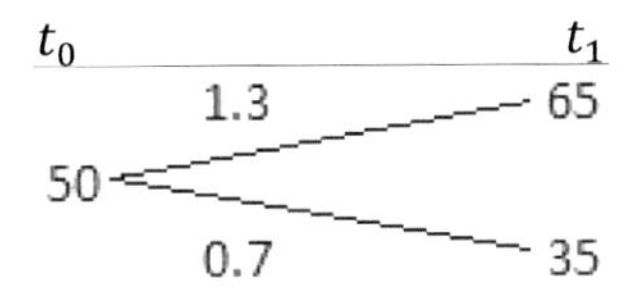

17 So if an option has 12 months to expiration $(T = 12)$ and there are two time increments $(n = 2)$, then the length of each time increment is $T/n = 12/2 = 6$ months.

18 Cox et al. (1979) took this approach.

19 See Endnote 4 in Section 4 for a refresher on standardized variables.

20 L'Hopital's rule is a helpful theorem for evaluating limits. It states $\lim_{x \to c} \frac{f(x)}{g(x)} = \lim_{x \to c} \frac{f'(x)}{g'(x)}$.

SECTION 6

Modern Portfolio Theory & CAPM

SOME ARGUE THAT MOVEMENTS in the price y of a security depends on some independent variable x, formally written as

$$y_t - y_{t-1} = \alpha + \beta(x_t - x_{t-1}) + e_t$$

$$\Delta y_t = \alpha + \beta \Delta x_t + e_t \tag{137}$$

where β is a coefficient describing the slope of the relationship, α is an intercept, and e_t (the error term) is the portion of Δy not explained by Δx.[1] What we mean when we say *independent variable* is simply that y does not itself effect x. The impact only runs in one direction, from x to y.

The parameters α and β are estimated by collecting samples of y and x and then performing regression analysis. Indexing the data available for y and x, we have y_t and x_t for times $t = \{1,\ldots,T\}$. Restating the model, we have

$$\Delta y_t = \hat{\alpha} + \hat{\beta}(\Delta x_t) + \varepsilon_t \tag{138}$$

The hats on the parameters represent the fact that they are estimates, based on sample data, of the population variables. The term ε_t is now referred to as the *residual*, which simply represents the difference between the observed data point Δy_t and the fitted data point $\Delta \hat{y}_t = \hat{\alpha} + \hat{\beta}(\Delta x_t)$.

DOI: 10.1201/9781032687650-6

6.1 LINEAR REGRESSION[2]

Estimating the parameters in Equation 138 is actually quite simple. One common approach to parameter estimation is Ordinary Least Squares (OLS). There are multiple ways to arrive at OLS estimates. Sections 6.1.1 and 6.1.2 cover two common approaches.

Note that OLS is a *method* of estimating model parameters. It is often useful, but not always the best approach. There are two desirable properties of parameter estimates:

1. The value of the parameter estimate is expected to equal the "true" parameter value.
2. Low deviance between the parameter estimate and the "true" parameter value.

There are two ways the first property may be met. If the expected value (mean) of a parameter estimate is equal to the true parameter value, the parameter estimate is said to be *unbiased*. Alternatively, if the parameter estimate converges in probability to the true parameter value as the sample size approaches infinity, the parameter estimate is said to be *consistent*.

The second property is judged by the variance, also known as the mean squared error, of the parameter estimate around the true parameter value. An unbiased estimate is said to be more *efficient* than an alternative unbiased estimate if it has a lower mean squared error.

If a linear estimator is both unbiased and has the lowest variance among all unbiased linear estimators, it is referred to as BLUE (**B**est **L**inear **U**nbiased **E**stimator).

To determine if an estimator produced by OLS is BLUE, one must determine whether a series of assumptions, known as the Gauss-Markov assumptions,[3] are met. If all Gauss-Markov assumptions are met, the OLS parameter estimate is BLUE. If certain subsets of the Gauss-Markov assumptions are met, the OLS parameter estimate may still be unbiased or consistent.

Unfortunately, even if a parameter estimate is BLUE, it may not be enough to perform statistical inference. We don't yet know the shape of the sampling distribution of the estimator. To perform proper inference, one must also assume that the error term (e_t from Equation 137) is normally distributed. The normal distribution is fully described by the first two moments (expected value and variance). If an estimator is BLUE, we would implicitly know its expected value and variance. Knowledge of the sampling distribution allows one to make probabilistic statements and perform hypothesis tests with respect to the true parameter value.

6.1.1 Method of Moments

To ease notation, we will drop the Δ symbol and just use y and x. While we are working with the change in y and x in Equation 138, including Δ might make the notation more difficult to follow. The Method of Moments relies upon two assumptions to arrive at OLS estimates of the regression parameters: one harmless and one crucial. Let's start by restating our regression equation.

$$y_t = \hat{\alpha} + \hat{\beta} x_t + \varepsilon_t$$

With the inclusion of the intercept $\hat{\alpha}$, it is harmless to assume that the average value of ε_t in the population is zero. In other words, $E(\varepsilon_t) = 0$. This assumption simply dictates where to concentrate the impact of unobserved factors. If $E(\varepsilon_t) = 0$ is not true, one could simply adjust the intercept to make this assumption true.

The second and more important assumption regards the relationship between ε_t and x_t. If some form of dependence between ε_t and x_t is present, linear or nonlinear, it may lead to problems with model interpretation and statistical inference. Accordingly, it is important that the average value of ε_t is independent of x_t. Formally, $E(\varepsilon_t \mid x_t) = E(\varepsilon_t)$.

When this latter assumption holds, ε_t is said to be *mean independent* of x. In conjunction with the first assumption, we have the zero conditional mean assumption $E(\varepsilon_t | x_t) = E(\varepsilon_t) = 0$.

With the zero conditional mean assumption in mind, we can now make two useful statements for uncovering the OLS estimates of the regression parameters:

$$E(\varepsilon_t) = 0$$

$$\text{cov}(x_t, \varepsilon_t) = E(x_t \varepsilon_t) = 0$$

The second statement follows from mean independence. Rewriting these statements, we have

$$E(\varepsilon_t) = E\left(y_t - \hat{\alpha} - \hat{\beta} x_t\right) = 0$$

and

$$E(x_t \varepsilon_t) = E\left[x_t\left(y_t - \hat{\alpha} - \hat{\beta} x_t\right)\right] = 0$$

These relations provide constraints for the joint probability distribution of x_t and y_t. For a given sample, we then have

$$E\left(y_t - \hat{\alpha} - \hat{\beta}x_t\right) = \frac{1}{T}\sum_{t=1}^{T}\left(y_t - \hat{\alpha} - \hat{\beta}x_t\right) = 0 \tag{139}$$

and

$$E\left[x_t\left(y_t - \hat{\alpha} - \hat{\beta}x_t\right)\right] = \frac{1}{T}\sum_{t=1}^{T}x_t\left(y_t - \hat{\alpha} - \hat{\beta}x_t\right) = 0 \tag{140}$$

Continuing to rewrite, for Equation 139 we have

$$\frac{1}{T}\sum_{t=1}^{T}\left(y_t - \hat{\alpha} - \hat{\beta}x_t\right) = \left(\frac{1}{T}\sum_{t=1}^{T}y_t\right) - \left(\hat{\alpha}\frac{1}{T}\sum_{t=1}^{T}1\right) - \left(\hat{\beta}\frac{1}{T}\sum_{t=1}^{T}x_t\right)$$

$$= \overline{y}_t - \hat{\alpha} - \hat{\beta}\overline{x}_t = 0$$

where the bar above the variable indicates the sample average. Solving for the intercept, we get

$$\hat{\alpha} = \overline{y}_t - \hat{\beta}\overline{x}_t \tag{141}$$

Plugging Equation 141 into Equation 140, we have

$$\frac{1}{T}\sum_{t=1}^{T}x_t\left(y_t - \hat{\alpha} - \hat{\beta}x_t\right) = \frac{1}{T}\sum_{t=1}^{T}x_t\left(y_t - \left(\overline{y}_t - \hat{\beta}\overline{x}_t\right) - \hat{\beta}x_t\right)$$

$$= \frac{1}{T}\sum_{t=1}^{n}x_t\left(y_t - \overline{y}_t + \hat{\beta}\overline{x}_t - \hat{\beta}x_t\right)$$

$$= \frac{1}{T}\sum_{t=1}^{T}x_t\left(y_t - \overline{y}_t\right) + \hat{\beta}\frac{1}{T}\sum_{t=1}^{T}x_t\left(\overline{x}_t - x_t\right) = 0$$

meaning

$$\sum_{t=1}^{T}x_t\left(y_t - \overline{y}_t\right) = \hat{\beta}\sum_{t=1}^{T}x_t\left(x_t - \overline{x}_t\right) \tag{142}$$

Let's explore $\sum_{t=1}^{T} x_t(x_t - \bar{x}_t)$.

$$\sum_{t=1}^{T} x_t(x_t - \bar{x}_t) = \sum_{t=1}^{T} x_t^2 - \sum_{t=1}^{T} x_t \bar{x}$$

$$= \sum_{t=1}^{T} x_t^2 - \sum_{t=1}^{T} x_t \frac{1}{T}\sum_{t=1}^{T} x_t$$

$$= \sum_{t=1}^{T} x_t^2 - T\frac{1}{T}\sum_{t=1}^{T} x_t \frac{1}{T}\sum_{t=1}^{T} x_t$$

$$= \sum_{t=1}^{T} x_t^2 - T(\bar{x})^2$$

Now consider $\sum_{t=1}^{T} (x_t - \bar{x}_t)(x_t - \bar{x}_t)$.

$$\sum_{t=1}^{T} (x_t - \bar{x}_t)(x_t - \bar{x}_t) = \sum_{t=1}^{T} x_t^2 - 2x_t\bar{x}_t + (\bar{x})^2$$

$$= \sum_{t=1}^{T} x_t^2 - \sum_{t=1}^{T} 2x_t\bar{x}_t + \sum_{t=1}^{T} (\bar{x})^2$$

$$= \sum_{t=1}^{T} x_t^2 - 2T\frac{1}{T}\sum_{t=1}^{T} x_t \frac{1}{T}\sum_{t=1}^{T} x_t + T(\bar{x})^2$$

$$= \sum_{t=1}^{T} x_t^2 - 2T(\bar{x})^2 + T(\bar{x})^2$$

$$= \sum_{t=1}^{T} x_t^2 - T(\bar{x})^2$$

We have just shown that $\sum_{t=1}^{T} x_t(x_t - \bar{x}_t) = \sum_{t=1}^{T}(x_t - \bar{x}_t)^2$. By the same proof, $\sum_{t=1}^{T} x_t(y_t - \bar{y}_t) = \sum_{t=1}^{T}(x_t - \bar{x})(y_t - \bar{y}_t)$. Substituting these values into Equation 142 gives us

$$\sum_{t=1}^{T}(x_t - \bar{x})(y_t - \bar{y}_t) = \hat{\beta}\sum_{t=1}^{T}(x_t - \bar{x}_t)^2$$

$$\hat{\beta} = \frac{\sum_{t=1}^{T}(x_t - \bar{x})(y_t - \bar{y}_t)}{\sum_{t=1}^{T}(x_t - \bar{x}_t)^2}$$

Multiplying the numerator and denominator by $\frac{1}{T-1}$, we get

$$\hat{\beta} = \frac{\text{cov}(x_t, y_t)}{\sigma_{x_t}^2}$$

This is the OLS solution for the slope parameter.

6.1.2 Sum of Squared Residuals

Next, we arrive at the OLS estimates by finding the parameter values that minimize the sum of squared residuals. We will drop the intercept α for now to help build intuition but will add it back later in this section. Restating the model whose parameters are to be estimated, we have

$$y_t = \hat{\beta}x_t + \varepsilon_t$$

Our goal is to minimize the sum of squared residuals.

$$\text{minimize} \sum_{t=1}^{T} \varepsilon_t^2$$

Since residuals represent the distance between actual and fitted values of y, minimizing the residuals will ensure fitted values are as close to observed values, on average, as possible. Why minimize the squared residuals? Well, one could minimize the sum of the absolute value of residuals or minimize the sum of residuals to an even higher power. The higher the power, the more weight is given to larger residuals in determining the best fit. Squared residuals are generally used because the math is easier than using absolute value, without giving too much weight to larger residuals.

Rewriting the sum of squared residuals, we have

$$\sum_{t=1}^{T}\varepsilon_t^2 = \sum_{t=1}^{T}\left(y_t - \hat{\beta}x_t\right)^2$$

$$= \sum_{t=1}^{T} y_t^2 - 2y_t\hat{\beta}x_t + \hat{\beta}^2 x_t^2$$

$$= \sum_{t=1}^{T} y_t^2 - \sum_{t=1}^{T} 2y_t\hat{\beta}x_t + \sum_{t=1}^{T}\hat{\beta}^2 x_t^2$$

To minimize this function, we will take the partial derivative with respect to $\hat{\beta}$, set this equal to zero, and then solve for $\hat{\beta}$.[4]

$$\frac{d}{d\hat{\beta}}\left[\sum_{t=1}^{T} y_t^2 - \sum_{t=1}^{T} 2y_t\hat{\beta}x_t + \sum_{t=1}^{T}\hat{\beta}^2 x_t^2\right] = \frac{d}{d\hat{\beta}}\sum_{t=1}^{T} y_t^2 - \frac{d}{d\hat{\beta}}\sum_{t=1}^{T} 2y_t\hat{\beta}x_t + \frac{d}{d\hat{\beta}}\sum_{t=1}^{T}\hat{\beta}^2 x_t^2$$

$$= 0 - \sum_{t=1}^{T} 2y_t x_t + \sum_{t=1}^{T} 2\hat{\beta}x_t^2 = 0$$

Rearranging, we have

$$\sum_{t=1}^{T} 2y_t x_t = \sum_{t=1}^{T} 2\hat{\beta}x_t^2$$

$$\sum_{t=1}^{T} y_t x_t = \hat{\beta}\sum_{t=1}^{T} x_t^2$$

$$\hat{\beta} = \frac{\sum_{t=1}^{T} y_t x_t}{\sum_{t=1}^{T} x_t^2}$$

Multiplying the numerator and denominator by $\frac{1}{T-1}$ would set the numerator equal to the sample covariance between variables x and y, and the denominator equal to the sample variance of variable x. This gives us the same OLS solution for the slope derived using the Method of Moments:

$$\hat{\beta} = \frac{\text{cov}(y_t, x_t)}{\sigma_{x_t}^2}$$

Now that we have the OLS estimate for the slope parameter without an intercept, let's add the intercept back in and repeat the exercise. We start with

$$y_t = \hat{\alpha} + \hat{\beta} x_t + \varepsilon_t$$

Rewriting the sum of squared residuals again, we have

$$\sum_{t=1}^{T} \varepsilon_t^2 = \sum_{t=1}^{T} \left(y_t - \hat{\alpha} - \hat{\beta} x_t \right)^2 \tag{143}$$

Since

$$\begin{aligned}\left(y_t - \hat{\alpha} - \hat{\beta} x_t \right)\left(y_t - \hat{\alpha} - \hat{\beta} x_t \right) &= y_t^2 - y_t \hat{\alpha} - y_t \hat{\beta} x_t - \hat{\alpha} y_t + \hat{\alpha}^2 \\ &\quad + \hat{\alpha}\hat{\beta} x_t - \hat{\beta} x_t y_t + \hat{\beta} x_t \hat{\alpha} + \hat{\beta}^2 x_t^2 \\ &= y_t^2 - 2 y_t \hat{\alpha} - 2 y_t \hat{\beta} x_t + 2\hat{\alpha}\hat{\beta} x_t + \hat{\alpha}^2 + \hat{\beta}^2 x_t^2\end{aligned}$$

we can rewrite Equation 143 as

$$\begin{aligned}&\sum_{t=1}^{T} y_t^2 - 2 y_t \hat{\alpha} - 2 y_t \hat{\beta} x_t + 2\hat{\alpha}\hat{\beta} x_t + \hat{\alpha}^2 + \hat{\beta}^2 x_t^2 \\ &= \sum_{t=1}^{T} y_t^2 - \sum_{t=1}^{T} 2 y_t \hat{\alpha} - \sum_{t=1}^{T} 2 y_t \hat{\beta} x_t + \sum_{t=1}^{T} 2\hat{\alpha}\hat{\beta} x_t + \sum_{t=1}^{T} \hat{\alpha}^2 + \sum_{t=1}^{T} \hat{\beta}^2 x_t^2\end{aligned} \tag{144}$$

To find $\hat{\alpha}$, we take the partial derivative with respect to $\hat{\alpha}$, set this equal to zero, and then solve for $\hat{\alpha}$

$$\frac{d}{d\hat{\alpha}}\left[\sum_{t=1}^{T} y_t^2 - \sum_{t=1}^{T} 2y_t\hat{\alpha} - \sum_{t=1}^{T} 2y_t\hat{\beta}x_t + \sum_{t=1}^{T} 2\hat{\alpha}\hat{\beta}x_t + \sum_{t=1}^{T} \hat{\alpha}^2 + \sum_{t=1}^{T} \hat{\beta}^2 x_t^2\right]$$

$$= 0 - \sum_{t=1}^{T} 2y_t - 0 + \sum_{t=1}^{T} 2\hat{\beta}x_t + \sum_{t=1}^{T} 2\hat{\alpha} + 0 = 0$$

Rearranging we get

$$\sum_{t=1}^{T} 2\hat{\beta}x_t + \sum_{t=1}^{T} 2\hat{\alpha} = \sum_{t=1}^{T} 2y_t$$

$$\sum_{t=1}^{T} \hat{\beta}x_t + \hat{\alpha}\sum_{t=1}^{T} 1 = \sum_{t=1}^{T} y_t$$

$$\hat{\alpha}T = \sum_{t=1}^{T} y_t - \sum_{t=1}^{T} \hat{\beta}x_t$$

$$\hat{\alpha} = \frac{1}{T}\sum_{t=1}^{T} y_t - \hat{\beta}\frac{1}{T}\sum_{t=1}^{T} x_t$$

$$= \overline{y_t} - \hat{\beta}\overline{x_t}$$

This matches the intercept estimate derived from Equation 141. To find the OLS estimate for the slope, we return to Equation 144 and perform the same procedure but take the partial derivative with respect to $\hat{\beta}$ instead, set this equal to zero, and then solve for $\hat{\beta}$:

$$\frac{d}{d\hat{\beta}}\left[\sum_{t=1}^{T} y_t^2 - \sum_{t=1}^{T} 2y_t\hat{\alpha} - \sum_{t=1}^{T} 2y_t\hat{\beta}x_t + \sum_{t=1}^{T} 2\hat{\alpha}\hat{\beta}x_t + \sum_{t=1}^{T} \hat{\alpha}^2 + \sum_{t=1}^{T} \hat{\beta}^2 x_t^2\right]$$

$$= 0 - 0 - \sum_{t=1}^{T} 2y_t x_t + \sum_{t=1}^{T} 2\hat{\alpha}x_t + 0 + \sum_{t=1}^{T} 2\hat{\beta}x_t^2 = 0$$

Rearranging, we have

$$\sum_{t=1}^{T} 2y_t x_t = \sum_{t=1}^{T} 2\hat{\alpha} x_t + \sum_{t=1}^{T} 2\hat{\beta} x_t^2$$

$$\sum_{t=1}^{T} y_t x_t = \sum_{t=1}^{T} \hat{\alpha} x_t + \hat{\beta} \sum_{t=1}^{T} x_t^2$$

$$\hat{\beta} = \frac{\sum_{t=1}^{T} y_t x_t - \sum_{t=1}^{T} \hat{\alpha} x_t}{\sum_{t=1}^{T} x_t^2}$$

$$= \frac{\sum_{t=1}^{T} \left(y_t x_t - \hat{\alpha} x_t \right)}{\sum_{t=1}^{T} x_t^2}$$

$$= \frac{\sum_{t=1}^{T} x_t \left(y_t - \hat{\alpha} \right)}{\sum_{t=1}^{T} x_t^2}$$

This should be intuitive. When an intercept is included, the slope must be set after offsetting the dependent variable with the intercept. Substituting $\bar{y} - \hat{\beta}\bar{x}$ for $\hat{\alpha}$ and rearranging results in the same OLS estimate

$$\hat{\beta} = \frac{\text{cov}\left(x_t, y_t\right)}{\sigma_{x_t}^2}$$

To prove this, we return to the beginning

$$= \sum_{t=1}^{T} \left(y_t - \hat{\alpha} - \hat{\beta} x_t \right)^2$$

$$= \sum_{t=1}^{T} \left(y_t - \overline{y_t} + \hat{\beta}\overline{x_t} - \hat{\beta} x_t \right)^2$$

$$= \sum_{t=1}^{T} \left(y_t - \overline{y_t} + \hat{\beta}\left(\overline{x_t} - x_t\right)\right)^2$$

$$= \sum_{t=1}^{T} y_t^2 - y_t \overline{y_t} + y_t \hat{\beta}\left(\overline{x_t} - x_t\right) - y_t \overline{y_t} + \overline{y_t}^2 - \overline{y_t}\hat{\beta}\left(\overline{x_t} - x_t\right)$$

$$+ y_t \hat{\beta}\left(\overline{x_t} - x_t\right) - \overline{y_t}\hat{\beta}\left(\overline{x_t} - x_t\right) + \hat{\beta}^2\left(\overline{x_t} - x_t\right)^2$$

$$= \sum_{t=1}^{T} y_t^2 - 2 y_t \overline{y_t} + 2 y_t \hat{\beta}\left(\overline{x_t} - x_t\right) + \overline{y_t}^2 - 2\overline{y_t}\hat{\beta}\left(\overline{x_t} - x_t\right) + \hat{\beta}^2\left(\overline{x_t} - x_t\right)^2$$

$$= \sum_{t=1}^{T} y_t^2 - \sum_{t=1}^{T} 2 y_t \overline{y_t} + 2\sum_{t=1}^{T} y_t \hat{\beta}\left(\overline{x_t} - x_t\right) + \sum_{t=1}^{T} \overline{y_t}^2 - 2\sum_{t=1}^{T} \overline{y_t}\hat{\beta}\left(\overline{x_t} - x_t\right)$$

$$+ \sum_{t=1}^{T} \hat{\beta}^2\left(\overline{x_t} - x_t\right)^2$$

Taking the partial derivative with respect to the slope gives us

$$\frac{d}{d\hat{\beta}}\left[\sum_{t=1}^{T} y_t^2 - \sum_{t=1}^{T} 2 y_t \overline{y_t} + 2\sum_{t=1}^{T} y_t \hat{\beta}\left(\overline{x_t} - x_t\right) + \sum_{t=1}^{T} \overline{y_t}^2 - 2\sum_{t=1}^{T} \overline{y_t}\hat{\beta}\left(\overline{x_t} - x_t\right) + \sum_{t=1}^{T} \hat{\beta}^2\left(\overline{x_t} - x_t\right)^2\right]$$

$$= \frac{d}{d\hat{\beta}}\sum_{t=1}^{T} y_t^2 - \frac{d}{d\hat{\beta}}\sum_{t=1}^{T} 2 y_t \overline{y_t} + 2\frac{d}{d\hat{\beta}}\sum_{t=1}^{T} y_t \hat{\beta}\left(\overline{x_t} - x_t\right)$$

$$+ \frac{d}{d\hat{\beta}}\sum_{t=1}^{T} \overline{y_t}^2 - 2\frac{d}{d\hat{\beta}}\sum_{t=1}^{T} \overline{y_t}\hat{\beta}\left(\overline{x_t} - x_t\right) + \frac{d}{d\hat{\beta}}\sum_{t=1}^{T} \hat{\beta}^2\left(\overline{x_t} - x_t\right)^2$$

$$= 0 - 0 + 2\sum_{t=1}^{T} y_t\left(\overline{x_t} - x_t\right) + 0 - 2\sum_{t=1}^{T} \overline{y_t}\left(\overline{x_t} - x_t\right) + 2\sum_{t=1}^{T} \hat{\beta}\left(\overline{x_t} - x_t\right)^2$$

Setting this equal to zero and rearranging gives us

$$2\sum_{t=1}^{T} y_t\left(\overline{x_t}-x_t\right)-2\sum_{t=1}^{T}\overline{y_t}\left(\overline{x_t}-x_t\right)+2\hat{\beta}\sum_{t=1}^{T}\left(\overline{x_t}-x_t\right)^2=0$$

$$\hat{\beta}\sum_{t=1}^{T}\left(\overline{x_t}-x_t\right)^2=\sum_{t=1}^{T}\overline{y_t}\left(\overline{x_t}-x_t\right)-\sum_{t=1}^{T}y_t\left(\overline{x_t}-x_t\right)$$

$$=\sum_{t=1}^{T}\overline{y_t}\left(\overline{x_t}-x_t\right)-y_t\left(\overline{x_t}-x_t\right)$$

$$=\sum_{t=1}^{T}\overline{y_t}\,\overline{x_t}-\overline{y_t}x_t-y_t\overline{x_t}+y_tx_t$$

$$=\sum_{t=1}^{T}\left(x_t-\overline{x}\right)\left(y_t-\overline{y}\right)$$

meaning

$$\hat{\beta}=\frac{\sum_{t=1}^{T}\left(x_t-\overline{x}\right)\left(y_t-\overline{y}\right)}{\sum_{t=1}^{T}\left(\overline{x_t}-x_t\right)^2}$$

$$=\frac{\operatorname{cov}\left(x_t,y_t\right)}{\sigma_{x_t}^2}$$

6.2 MODERN PORTFOLIO THEORY

Modern Portfolio Theory ("MPT") provides a framework for understanding portfolio risk. Before applying what we learned in Section 6.1 to securities markets to derive the Capital Asset Pricing Model (CAPM), we must understand risk within a portfolio context. We will then see how the CAPM is a simple extension of MPT.

6.2.1 Primer on Risk Proxies[5]

Since we are diving into a framework for dealing with multiple assets, it's prudent to refresh our knowledge of probability distributions. To do this, we introduce the concept of *moments*. The moments of a function provide quantitative representations of the location, scale, and shape of the function's probability distribution. There are many types of moments, but for our purposes, we focus on four: mean, variance, skewness, and kurtosis.

Generally speaking, one may think of the nth moment of the continuous function $f(X)$ around some constant c as

$$m_n = E[X-c]^n$$

$$= \int_{-\infty}^{\infty} (x-c)^n f(x)dx$$

When calculating the mean, also known as the first moment (meaning $n=1$), we will set $c=0$. In this case, we have

$$m_1 = \int_{-\infty}^{\infty} xf(x)dx$$

This should be familiar, as it is the textbook formula for the mean of a continuous random variable (see Section 1.1). With c set to zero, the mean is referred to here as a *raw* moment, in that it is not calculated relative to any other value. The mean reflects an expectation. Say X is the return of some security. If one holds a security long enough to experience returns spanning the entire probability distribution, the average return will equal the mean. Visualizing the mean using a fictitious probability distribution, we restate Figure 4.10:[6]

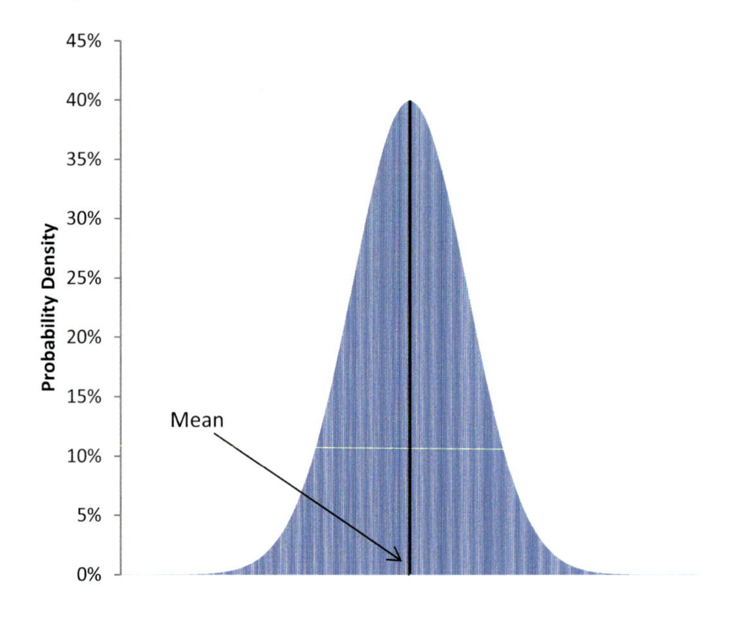

When calculating the variance, also known as the second central moment (meaning $n = 2$), we no longer set $c = 0$. Rather, we will set $c = m_1$. In this case, we have

$$m_2 = \int_{-\infty}^{\infty} (x - m_1)^2 f(x)dx$$

With c set to the mean, variance is referred to here as a *central* moment. Variance represents dispersion from the mean. It depicts how close or far the data tends to be relative to the mean. In the context of a security's return, variance indicates whether returns tend to cluster near the mean or vary far above and below the mean. Figure 6.1 provides a visualization of variance using fictitious probability distributions:

The blue distribution is the same as the prior visualization of the mean in Figure 4.10. The green distribution has a lower variance, and the red distribution has a higher variance.

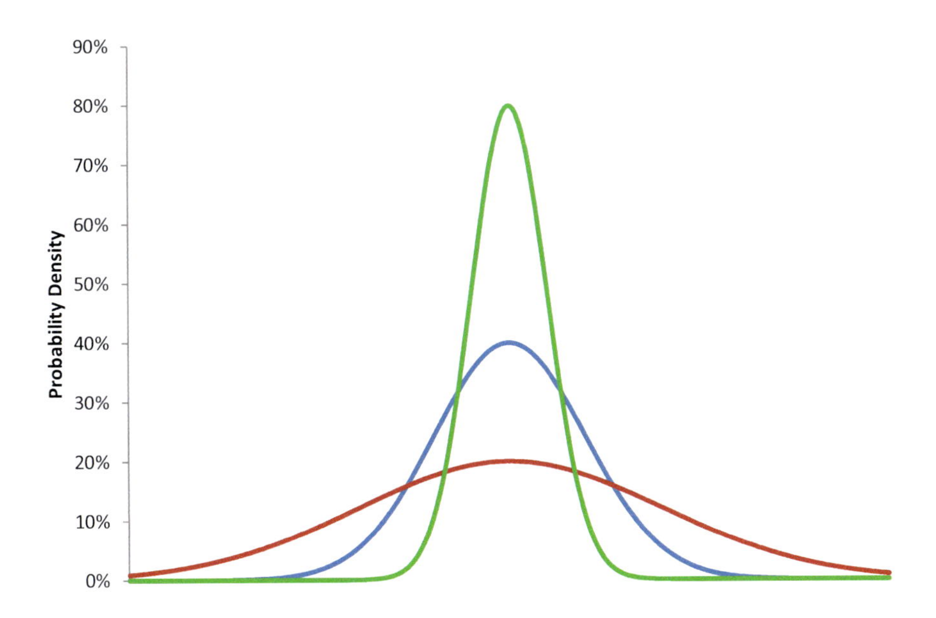

FIGURE 6.1 Visualization of variance.

Since data comes in all shapes and sizes, comparing one distribution to another using moments requires moments to be stated in terms of some common unit. To do this, we introduce a *standardized* moment:

$$\frac{m_n}{\sigma^n} = \frac{E[X-c]^n}{E\left[(X-c)^2\right]^{\frac{n}{2}}}$$

Note that while the mean is a raw moment, the variance is defined in terms of the mean and higher [standardized] moments are defined in terms of the variance. By standardizing higher moments, moments may be better compared across distributions. We operate under the standardized moment framework when calculating skewness, also known as the third central moment (meaning $n=3$). In this case, we have

$$\frac{m_3}{\sigma^3} = \frac{E[X-m_1]^3}{E\left[(X-m_1)^2\right]^{\frac{3}{2}}}$$

$$= \frac{\int_{-\infty}^{\infty}(x-m_1)^3 f(x)dx}{\left[\int_{-\infty}^{\infty}(x-m_1)^2 f(x)dx\right]^{\frac{3}{2}}}$$

Skewness gives an impression of asymmetry in the return distribution. Figure 6.2 provides a visualization of skewness using fictitious probability distributions:

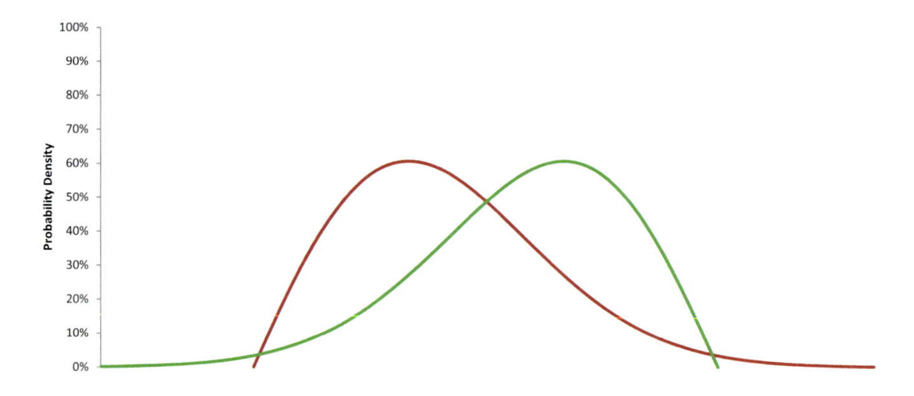

FIGURE 6.2 Visualization of skew.

While the distributions in the prior visualization of variance all have zero skew, we see here the green distribution with negative skew and the red distribution with positive skew. Due to the non-zero skew, the mean no longer equals the median and mode. For the green distribution, the mean is less than the median, and the median is less than the mode. The opposite is true of the red distribution.

We also operate under the standardized moment framework when calculating kurtosis, known as the fourth moment (meaning $n=4$). In this case, we have

$$\frac{m_4}{\sigma^4} = \frac{\int_{-\infty}^{\infty} (x-m_1)^4 f(x)dx}{\left[\int_{-\infty}^{\infty} (x-m_1)^2 f(x)dx\right]^2}$$

Kurtosis quantifies how much data lives in the tails of the distribution. On the flip side, it details how concentrated data is near the mean. Figure 6.3 provides a visualization of kurtosis using fictitious probability distributions:

Focus here on two places: the peaks of the distributions and the tails. The blue distribution is the same as the prior visualization of the mean in Figure 4.10. The red distribution has the lowest peak, while the green

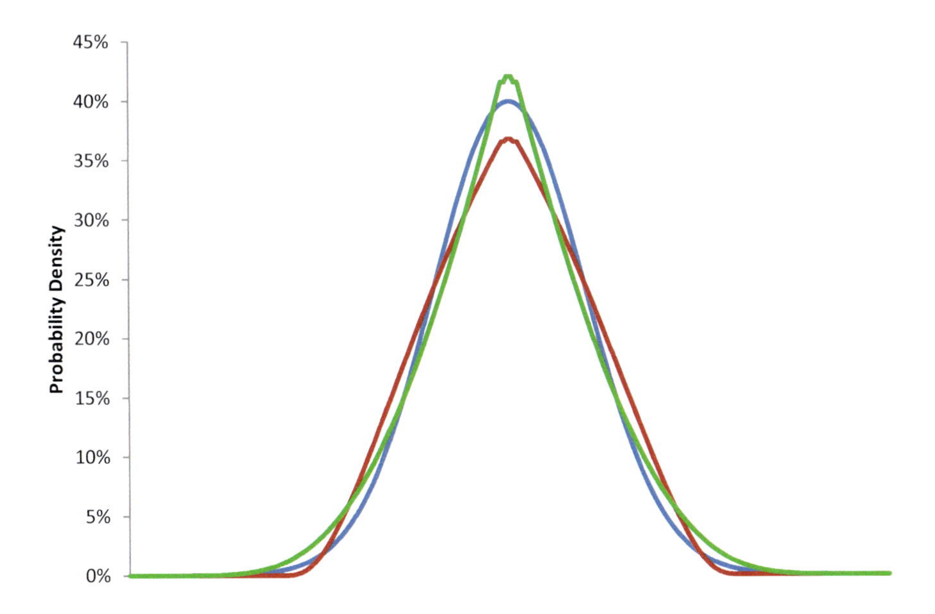

FIGURE 6.3 Visualization of kurtosis.

distribution has the highest peak. Notice also that the red distribution's tails decline fastest toward zero, while the green distribution's tails take the longest to decline toward zero. The red distribution has less kurtosis than the blue distribution, while the green distribution has higher kurtosis than the blue distribution.

Now consider the relation of the moments to each other. The mean, a raw moment, represents an expectation. All other moments, central and standardized, describe the shape of the data around the mean. If we experience the entire population of data, spanning the whole distribution, then we are guaranteed to experience the mean on average. If, however, we live in a finite world, where we experience only a sample from the distribution, then the sample average we experience might depend heavily on the shape of the distribution around the mean.

Accordingly, variance, skewness, and kurtosis might be viewed as proxies for *risk*. They describe ways in which sample properties might depart from population expectations. In the context of security returns, these risk proxies describe how the returns we experience over a finite period might stray from expected returns.

Each security has its own unique statistical moments that describe its returns over time. Building a portfolio of multiple securities will result in a portfolio return distribution that reflects a blend of the moments of the component securities. Since one's goal, generally, is to maximize expected return while minimizing risk, we might visualize the most effective portfolio construction strategy using a Pareto frontier.

A Pareto frontier is simply a curve that describes a tradeoff. Since we just discussed a tradeoff between expected return and risk, consider the Pareto frontier in Figure 6.4 in which we use variance as the proxy for risk:

Each blue dot represents the tradeoff between expected return and risk for some unique portfolio of securities. The Pareto frontier is depicted by the black line. Focus on the two portfolios with an expected return of 5%. One portfolio has a variance of 20%, while the other has a variance of 30%. Since the goal is to maximize return and minimize risk, the portfolio with 20% variance dominates the portfolio with 30% variance. Hence the portfolio with 20% variance falls on the black line. Similarly, consider the two portfolios with 60% variance. The portfolio with 8% expected return dominates the other with 7.5%, and so once again the dominant portfolio falls on the black line. The Pareto frontier therefore represents the set of all portfolios that provide maximal return for given levels of risk. Equivalently, it is the set of all portfolios that minimizes risk for given levels of return.

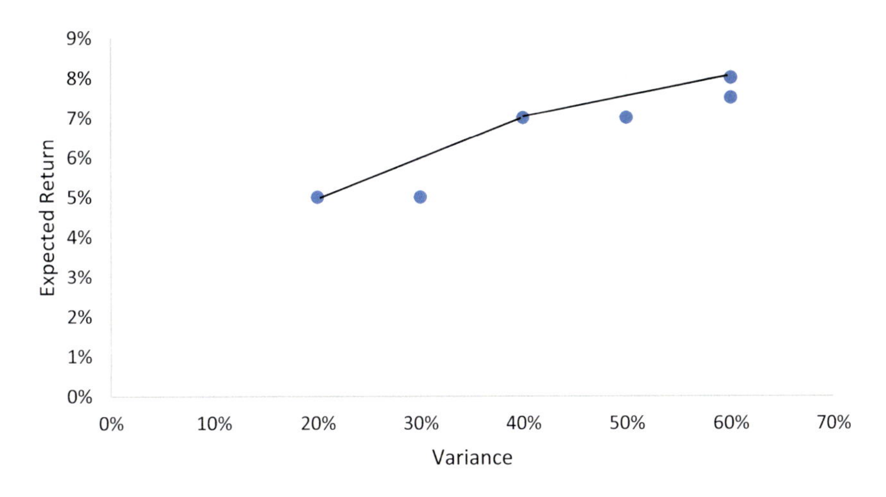

FIGURE 6.4 Pareto frontier.

6.2.2 Portfolio Context[7]

Now that we understand moments and the Pareto frontier, we can begin to understand how MPT defines rational portfolio construction.

MPT starts with the risk-return tradeoff. For a given expected return, we assume that investors prefer the less risky portfolio (i.e., they prefer the portfolio on the Pareto frontier, rather than a portfolio below the Pareto frontier). This amounts to assuming that investors are risk-averse.

CONCEPT REFRESHER 6.1: RISK AVERSION

Risk aversion refers to an individual's stance on the risk-return tradeoff. Specifically, it details how an individual would choose between risky alternatives.

Risk aversion is best demonstrated with an example. Consider a bet where one has a 50% chance of winning \$100 and a 50% chance of winning \$0. The expected value of taking this bet, of course, is \$50 $\left((50\% \times \$100) + (50\% \times \$0) = \$50\right)$. Say one could choose between taking this bet (Option A) and receiving the expected value outright (Option B). Which option should they choose?

The answer to this question reveals one's categorical preference when it comes to risk. The following table details the category-choice pairs:

Category	Choice
Risk-averse	Option B
Risk-neutral	Indifferent
Risk-tolerant	Option A

The risk-averse individual would rather take the expected value of the bet and avoid the risk completely. The only way they would accept risk is if it also carries a higher expected return. The risk-tolerant individual seeks risk, preferring to accept the risk so that they might have a chance at receiving the \$100 payout. The risk-neutral individual is indifferent between the two choices, having no preference one way or the other.

Next, we must choose the appropriate proxy for risk. While there are many ways to quantify risk, let's focus on the moments introduced in Section 6.2.1. If we assume that security prices are lognormally distributed (i.e., returns are normally distributed), then the choice of risk proxy becomes simple. Normally distributed random variables have constant skew and kurtosis. Specifically, zero skew and kurtosis of three. If one chooses kurtosis as the relevant risk measure when all securities are assumed to be normally distributed, then every security would have the exact same measure of risk! The exercise of constructing the best portfolio would simply amount to investing all funds in a single security with the highest expected return.

This makes skew, kurtosis, and any other metric that's constant across all securities inappropriate as a risk measure. Normally distributed variables are parameterized only by their mean and variance. Accordingly, the only viable proxy for risk under the normality assumption is variance. It follows that MPT relies on variance as the proxy for risk, as was used in the visualization of the Pareto frontier in Figure 6.4.

The goal now is to find the portfolios that maximize expected return for given levels of variance. However, we will make a slight modification. Variance is reported in squared units. This isn't always intuitive. It is common to take the square root of variance to align the risk units with that of the data. The square root of variance is referred to as the *standard deviation*.

Let's start by defining the expected return and standard deviation of a portfolio. For ease, we will consider a two-asset portfolio.

Portfolio variance σ^2 is defined

$$\sigma^2 = E\left[R_p - \mu\right]^2 \tag{145}$$

where $\mu = E\left(R_p\right)$ is the expected return of portfolio p. Putting this in the context of a portfolio, let's start by defining R_p. Since there are only two assets in the portfolio, assets 1 and 2, we have

$$R_p = w_1 R_1 + \left(1 - w_1\right) R_2$$

where w_i and R_i are the weight and return for asset i held in the portfolio. Taking expectation gives us

$$\begin{aligned} E(R_p) = \mu = E\left[w_1 R_1 + (1-w_1)R_2\right] \\ = w_1 E(R_1) + (1-w_1)E(R_2) \\ = w_1\mu_1 + (1-w_1)\mu_2 \end{aligned} \tag{146}$$

where μ_i is the expected (average) return for asset i. With this in mind, we can rewrite the variance from Equation 145 as

$$\begin{aligned} E\left[R_p - \mu_p\right]^2 &= E\left[w_1 R_1 + (1-w_1)R_2 - \{w_1\mu_1 + (1-w_1)\mu_2\}\right]^2 \\ &= E\left[w_1 R_1 - w_1\mu_1 + (1-w_1)R_2 - (1-w_1)\mu_2\right]^2 \\ &= E\left[w_1(R_1-\mu_1) + (1-w_1)(R_2-\mu_2)\right]^2 \\ &= E\begin{bmatrix} w_1^2(R_1-\mu_1)^2 + 2w_1(1-w_1)(R_1-\mu_1)(R_2-\mu_2) \\ +(1-w_1)^2(R_2-\mu_2)^2 \end{bmatrix} \\ &= E\left[w_1^2(R_1-\mu_1)^2\right] + E\left[2w_1(1-w_1)(R_1-\mu_1)(R_2-\mu_2)\right] \\ &\quad + E\left[(1-w_1)^2(R_2-\mu_2)^2\right] \\ &= w_1^2 E\left[(R_1-\mu_1)^2\right] + 2w_1(1-w_1)E\left[(R_1-\mu_1)(R_2-\mu_2)\right] \\ &\quad + (1-w_1)^2 E\left[(R_2-\mu_2)^2\right] \end{aligned} \tag{147}$$

Noting that $E\left[R_i - \mu_i\right]^2 = \sigma_i^2$ is the variance for asset i and $E\left[(R_i-\mu_i)(R_j-\mu_j)\right] = \text{cov}_{i,j}$ is the covariance between assets i and j, Equation 147 can be rewritten as

$$w_1^2\sigma_1^2 + 2w_1(1-w_1)\text{cov}_{1,2} + (1-w_1)^2\sigma_2^2$$

Rearranging and remembering that $\text{cov}_{i,j} = \sigma_i\sigma_j\rho_{i,j}$ where $\rho_{i,j}$ is the Pearson correlation coefficient between assets i and j, we recover the variance for the portfolio[8]:

$$\sigma_{\text{port}}^2 = w_1^2\sigma_1^2 + (1-w_1)^2\sigma_2^2 + 2w_1(1-w_1)\sigma_1\sigma_2\rho_{1,2}$$

Summarizing, we have expected portfolio return

$$w_1\mu_1 + (1-w_1)\mu_2$$

and portfolio standard deviation

$$\sqrt{w_1^2\sigma_1^2 + (1-w_1)^2\sigma_2^2 + 2w_1(1-w_1)\sigma_1\sigma_2\rho_{1,2}} \tag{148}$$

The final step is to select weights w_1 and w_2 that minimize the portfolio standard deviation for given levels of portfolio expected return.[9]

How does this look in practice? Let's provide a visual with a simple example. Take two assets.

1. Asset A:
   - Expected Return = 5%
   - Standard Deviation of Return = 10%
2. Asset B:
   - Expected Return = 15%
   - Standard Deviation of Return = 25%

Figure 6.5 details the *investment opportunity curves* for assets A and B. Three curves are shown, each under a different correlation assumption. Different points across any particular investment opportunity curve represent a unique weighting scheme between the two assets.

Consider first the green investment opportunity curve. This, in fact, is not a curve at all, but a straight line connecting the two possible portfolios. On the far left of the green line, we have a portfolio consisting 100% of asset A. Moving from left to right, the weight of asset A decreases and the weight of asset B increases until reaching the far right, where the weight of asset B reaches 100%. Since the correlation between assets A and B equals one on the green line, there is no diversification benefit.

Consider next the red and blue investment opportunity curves. At the ends, they converge with the green line, since having a portfolio that

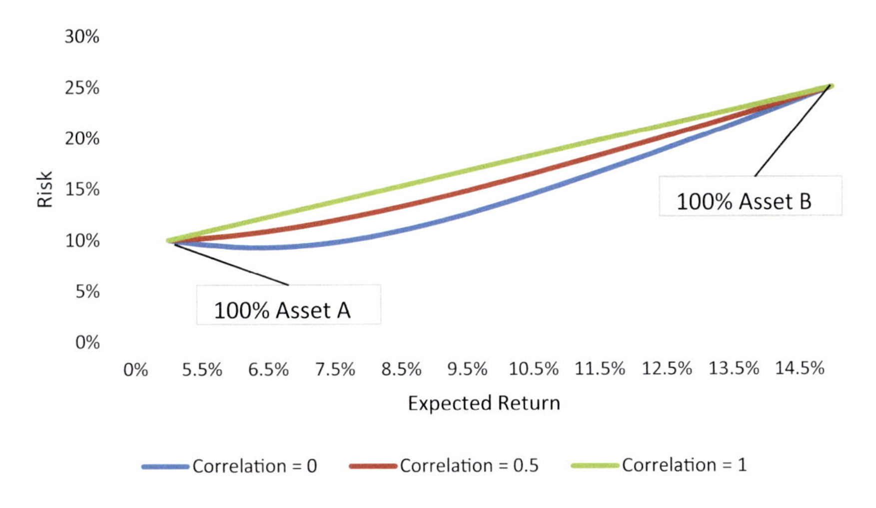

FIGURE 6.5 Investment opportunity curve (diversification benefit).

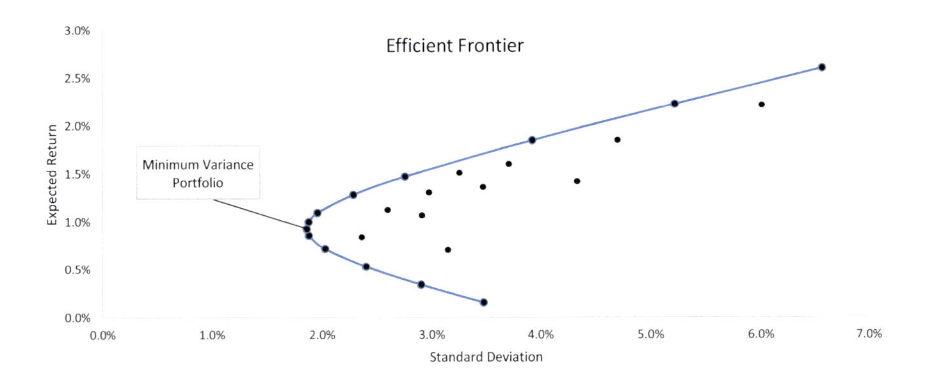

FIGURE 6.6 Efficient frontier.

consists only of one asset, by definition, has no diversification benefit. As the weights become more balanced, the red and blue curves offer a lower level of risk for each level of expected return. This is a diversification benefit, and as the correlation approaches zero, the diversification benefit increases. Once the correlation reaches −1, the diversification benefit is maximized for any given expected return.

Applying this across all investable assets, rather than just two, results in an efficient frontier for market portfolios. The efficient frontier for market portfolios is depicted in Figure 6.6:

Each of the black dots represents a unique portfolio. These portfolios are constructed by changing the weights applied to the whole universe of investable assets. All the portfolios on the blue line above the minimum variance portfolio, termed *efficient portfolios*, make up the efficient frontier. Portfolios below the minimum variance portfolio on the blue line, as well as those not on the blue line at all, are dominated by a portfolio on the efficient frontier.

There are now two interesting directions to take this. Consider creating new two-asset portfolios:

1. An efficient portfolio mixed with a risk-free asset.
2. An efficient portfolio mixed with a risky asset, not on the efficient frontier.

The first portfolio leads to the existence of the Capital Market Line (CML). The second portfolio, in conjunction with the CML, leads to the existence of the CAPM. The next two sections explain this in more detail.

6.2.3 Capital Market Line[10]

Now that we understand how to construct an efficient frontier, we can go a step further. We now have an additional portfolio construction choice: how should we allocate funds between a risky portfolio on the efficient frontier and a risk-free asset? To answer this question, we consider a new two-asset portfolio consisting of an efficient portfolio and a risk-free asset. First, we introduce the Sharpe ratio.

The Sharpe ratio λ is a simple metric used to quantify the risk-return tradeoff. Formally, the Sharpe ratio for asset i is written as

$$\lambda_i = \frac{E\left(R_i - r_f\right)}{\sigma_i}$$

where R_i is the return on asset i, r_f is the return on the risk-free asset, and σ_i is the standard deviation of return on asset i. This should be somewhat intuitive. The numerator represents the return of asset i in excess of the risk-free return, while the denominator is a proxy for risk. The Sharpe ratio therefore represents excess return per unit of risk. A high Sharpe ratio is implicitly preferred to a low Sharpe ratio, assuming investors are risk-averse and care only about the expected return and standard deviation of return of assets.[11]

Let's switch to thinking about the risk-return relationship of a two-asset portfolio. Recall from Figure 6.5 that one receives a diversification benefit by creating portfolios of assets that are not perfectly correlated. But what happens if one of the two assets is risk-free (i.e., zero standard deviation of return)? If we assume the second asset in a two-asset portfolio is risk-free, then the portfolio standard deviation σ_p formula simplifies to

$$\sigma_p = \sqrt{w_1^2\sigma_1^2 + (1-w_1)^2\sigma_2^2 + 2w_1(1-w_1)\sigma_1\sigma_2\rho_{1,2}}$$

$$= \sqrt{w_1^2\sigma_1^2 + (1-w_1)^2 0^2 + 2w_1(1-w_1)\sigma_1 0 \rho_{1,2}}$$

$$= \sqrt{w_1^2\sigma_1^2}$$

$$= w_1\sigma_1$$

The portfolio standard deviation is now just a scalar of the risky asset standard deviation, where the scalar is the risky asset's weight! The correlation no longer matters. This has the effect of turning the investment opportunity curve into a straight line. Restating Figure 6.5 for a two-asset portfolio wherein one asset is risk-free, we get Figure 6.7:

Notice that the curvature is gone. There is no longer a diversification benefit.

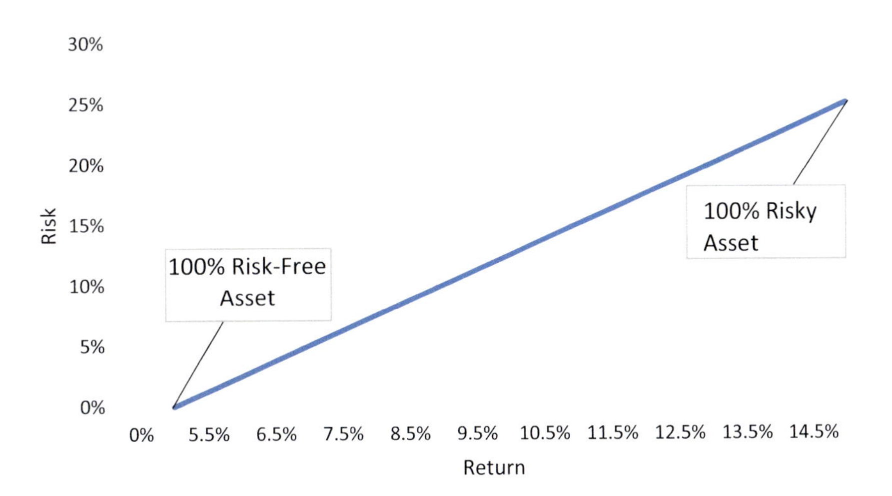

FIGURE 6.7 Investment opportunity curve with risk-free asset.

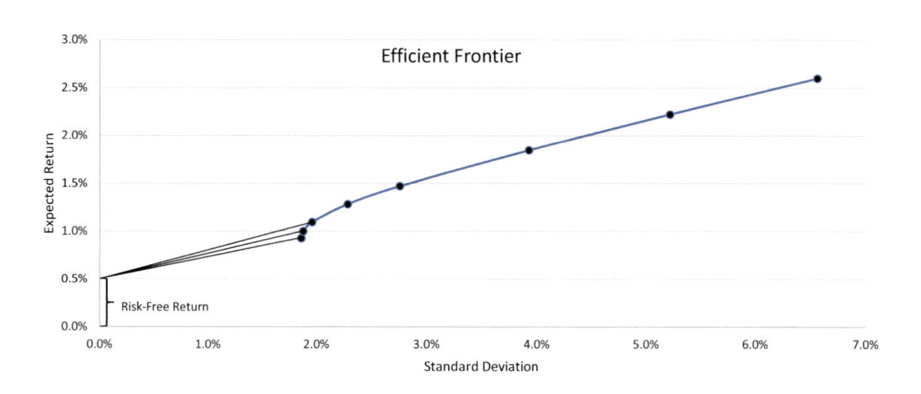

FIGURE 6.8 Investment opportunity curves intersecting the efficient frontier.

Next, let's visualize this new two-asset portfolio in the context of the efficient frontier. The risky asset might be any efficient portfolio from the efficient frontier. We assume the risk-free asset has an expected return of 0.5%. Figure 6.8 illustrates this scenario:

All inefficient portfolios have been deleted in Figure 6.8. Three different investment opportunity curves are shown by black lines, each representing portfolio combinations using the risk-free asset and some portfolios on the efficient frontier. If 100% weight is given to the risk-free asset, the portfolio characteristics will land at the far left of the three investment opportunity curves. If 100% weight is given to the risky asset, the portfolio will land on the black dot at the far right of one of the investment opportunity curves. Balancing weights will land you somewhere middling an investment opportunity curve.

Take a moment to calculate the Sharpe ratio λ_p for our two-asset portfolio p consisting of the risk-free asset with return r_f and a risky asset from the efficient frontier with return R_e and weight w_e. We already know the following:

$$E\left(R_p\right) = w_e E\left(R_e\right) + \left(1 - w_e\right) E\left(r_f\right)$$

$$\sigma_p = w_e \sigma_e$$

The Sharpe ratio is then

$$
\begin{aligned}
\lambda_p &= \frac{E\left(R_p - r_f\right)}{\sigma_p} \\
&= \frac{w_e E(R_e) + (1 - w_e) E\left(r_f\right) - E\left(r_f\right)}{w_e \sigma_e} \\
&= \frac{w_e E(R_e) + E\left(r_f\right) - w_e E\left(r_f\right) - E\left(r_f\right)}{w_e \sigma_e} \\
&= \frac{w_e E(R_e) - w_e E\left(r_f\right)}{w_e \sigma_e} \\
&= \frac{E(R_e) - r_f}{\sigma_e}
\end{aligned}
\tag{149}
$$

The weights have completely dropped out! The Sharpe ratio of the two-asset portfolio has reduced to the Sharpe ratio of the efficient portfolio. This means the Sharpe ratio is constant across an investment opportunity curve. Looking back at Figure 6.8, while the Sharpe ratio is different from one investment opportunity curve to another, the Sharpe ratio on any particular investment opportunity curve is constant. If one's preference is to attain a higher Sharpe ratio, then the investment opportunity curves may be referred to as *indifference curves*, since changing weights won't affect the Sharpe ratio once an efficient portfolio is chosen. Only choosing a new efficient portfolio will affect the Sharpe ratio.

If we assume that investors can borrow or lend at a risk-free rate, then we can extend the indifference curves beyond the risky portfolios on the efficient frontier. In other words, we could achieve, say, 150% weight to a risky portfolio by borrowing an additional 50% of our initial funds at the risk-free rate and then investing them in the risky portfolio.[12] Figure 6.9 illustrates this.

Again, each indifference curve represents a portfolio consisting of the risk-free asset and one of the efficient portfolios. The indifference curves extend beyond the risky portfolios on the efficient frontier because of the ability to leverage the investment. The new black dots at the end of the black lines represent these levered portfolios.

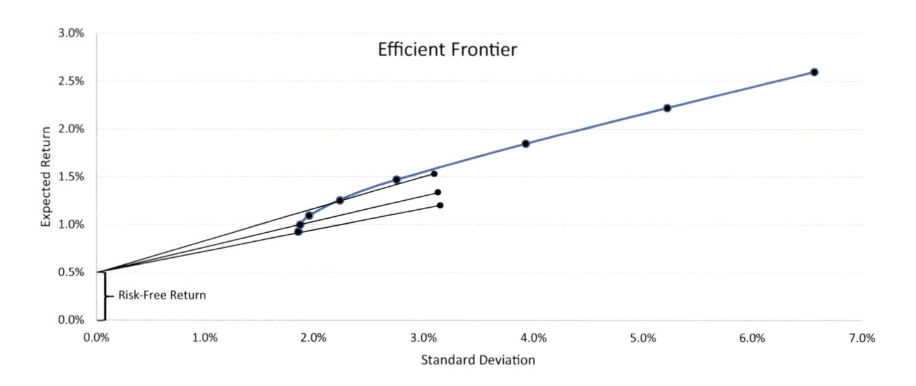

FIGURE 6.9 Levered indifference curves intersecting the efficient frontier.

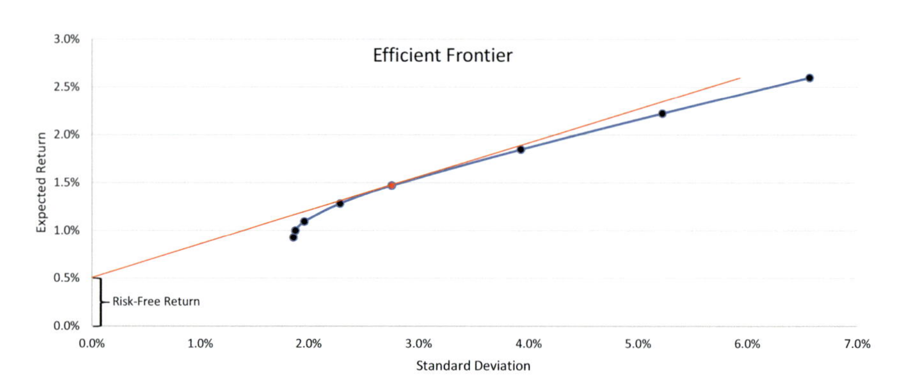

FIGURE 6.10 The capital market line.

But which portfolio on the efficient frontier should we choose as the risky asset to mix with the risk-free asset? In other words, which indifference curve should we choose? Well, we know two things: (1) each indifference curve has a constant Sharpe ratio and (2) the portfolios on the efficient frontier dominate those below the efficient frontier. Notice Figure 6.9. The levered portfolios lie below the efficient frontier! If the levered portfolios are suboptimal, then so are *all* the possible two-asset portfolios that lie on their corresponding indifference curve. There is only one point on the efficient frontier whose entire set of possible levered portfolios would never lie below the efficient frontier. This portfolio is referred to as the *market portfolio* (also known as the *tangency portfolio*), and is the portfolio found where an indifference curve is tangent to the efficient frontier. We see this in Figure 6.10.

The red line is the CML. The CML is a special indifference curve representing the set of two-asset portfolios consisting only of the risk-free asset and the market portfolio. The red dot that it crosses at the point of tangency with the efficient frontier is the market portfolio. This is the portfolio on the efficient frontier with the highest Sharpe ratio.[13] How do we know this? Well, we know from Equation 149 that the Sharpe ratio λ_p of our two-asset portfolio is equal to the Sharpe ratio λ_e of the risky asset. Setting these equal gives us

$$\frac{E(R_p)-r_f}{\sigma_p}=\frac{E(R_e)-r_f}{\sigma_e}$$

$$E(R_p)-r_f=\frac{E(R_e)-r_f}{\sigma_e}\sigma_p \tag{150}$$

$$E(R_p)=r_f+\frac{E(R_e)-r_f}{\sigma_e}\sigma_p$$

$$E(R_p)=r_f+\lambda_e\sigma_p$$

The slope of the indifference curve is the Sharpe ratio of the risky asset! If the indifference curve with the highest slope is chosen, then the Sharpe ratio will be maximized. As you might guess, the indifference curve with the highest slope will be that which is tangent to the efficient frontier. Any indifference curve with a higher slope is considered unattainable since it lies above the efficient frontier. Specifying the market portfolio as a risky asset, Equation 150 becomes

$$E(R_p)=r_f+\frac{E(R_m)-r_f}{\sigma_m}\sigma_p \tag{151}$$

where the subscript m denotes the market portfolio and $\frac{E(R_m)-r_f}{\sigma_m}=\lambda_m$ is the Sharpe ratio for the market portfolio.

The market portfolio is a fictitious construct. For one single market portfolio to exist, we must add the following assumptions to those already introduced:

1. *All investors have identical estimates of expected return, standard deviation of return, and correlations between returns for all investable assets.*
2. *All investors assess returns over a single, identical period.*
3. *Investment decisions are not influenced by taxes or transaction costs.*

These three assumptions ensure that all investors assess available data in the exact same way and don't come to alternative conclusions about investable assets. In other words, this ensures that all investors come to the same Sharpe ratio for every individual asset and portfolio of assets. Without these assumptions, investors might come to a different CML and a different tangency portfolio.

The CML should be somewhat intuitive – higher risk begets higher expected return. Two market-based prices emerge here:

- *Time Preference*: consider this the price of time. It represents the return attributable to time without any risk of loss and is illustrated by the risk-free return.
- *Risk Premium*: this is the price of risk. It represents the expected return in excess of the risk-free rate and is measured by the slope of the CML.

6.2.4 Capital Asset Pricing Model[14]

Sections 6.1 and 6.2.1–6.2.3 provide context for the model introduced in this section. The CAPM is generally the first model of asset returns that students learn in finance courses. It gives users a tool to calculate the expected return for any financial asset, given its systemic risk profile. This is, of course, conditional on all of the assumptions discussed since the beginning of Section 6.2.

Loosely speaking, our goal is to use what we know about portfolios on the efficient frontier to learn about the expected return of an individual asset (which may be partially included in an efficient portfolio). Restating what we know:

- From Equation 146, the expected return $E(R_p)$ for a two-asset portfolio is

$$E(R_p) = w_A E(R_A) + (1 - w_A) E(R_B) \tag{152}$$

  where w is the weight and subscripts A and B index the two assets, A and B, respectively.

- From Equation 148, the standard deviation σ_p for a two-asset portfolio is

$$\sigma_p = \sqrt{w_A^2 \sigma_A^2 + (1 - w_A)^2 \sigma_B^2 + 2 w_A (1 - w_B) \sigma_A \sigma_B \rho_{A,B}} \tag{153}$$

- From Equation 151, the equation for the CML is

$$E(R_p) = r_f + \frac{E(r_m) - r_f}{\sigma_m}\sigma_p \tag{154}$$

where r_f is the risk-free rate and subscript m refers to the market portfolio.

Now let's consider another new two-asset portfolio p. The first asset will be some risky asset i. The second asset will be the market portfolio (still denoted with subscript m). Restating the expected return and standard deviation from Equations 152 and 153, respectively, in terms of this new two-asset portfolio, we have

$$E(R_p) = w_i E(R_i) + (1 - w_i) E(R_m) \tag{155}$$

$$\sigma_p = \sqrt{w_i^2 \sigma_i^2 + (1 - w_i)^2 \sigma_m^2 + 2w_i(1 - w_i)\sigma_i \sigma_m \rho_{i,m}} \tag{156}$$

Figure 6.11 provides a visualization of this two-asset portfolio.

The black dot labeled **i** represents the risky asset and the red dot still represents the market portfolio. The black curve connecting them is their investment opportunity curve. The black dot labeled **i'** represents the portfolio where asset i is fully shorted (i.e., weight $w_i = -100\%$). Note that at $w_i = 0\%$, the two-asset portfolio effectively becomes the market portfolio, meaning the investment opportunity curve intersects with the CML.

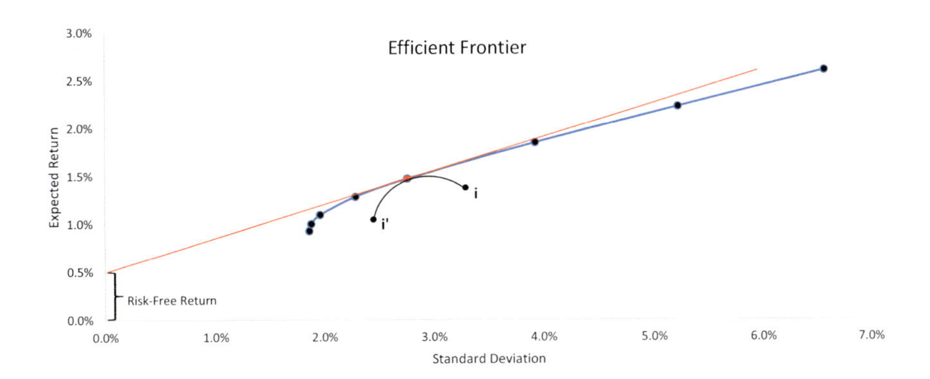

FIGURE 6.11 CAPM visualization.

More specifically, at $w_i = 0\%$ the slope of the investment opportunity curve is equal to the slope of the CML.

To find the slope of the investment opportunity curve at $w_i = 0\%$, we must first find the first-order sensitivities of Equations 155 and 156 to changes in the weight to asset i. For Equation 155, we have

$$\begin{aligned}\frac{\partial E(R_p)}{\partial w_i} &= \frac{\partial}{\partial w_i} w_i E(R_i) + (1 - w_i) E(R_m) \\ &= E(R_i) - E(R_m)\end{aligned} \tag{157}$$

For Equation 156, we have

$$\begin{aligned}\frac{\partial \sigma_p}{\partial w_i} &= \frac{\partial}{\partial w_i} \sqrt{w_i^2 \sigma_i^2 + (1 - w_i)^2 \sigma_m^2 + 2 w_i (1 - w_i) \sigma_i \sigma_m \rho_{i,m}} \\ &= \frac{\partial}{\partial w_i} \left(w_i^2 \sigma_i^2 + (1 - w_i)^2 \sigma_m^2 + 2 w_i (1 - w_i) \sigma_i \sigma_m \rho_{i,m} \right)^{\frac{1}{2}} \\ &= \frac{\partial}{\partial w_i} \left(w_i^2 \sigma_i^2 + \sigma_m^2 - 2 w_i \sigma_m^2 + w_i^2 \sigma_m^2 + 2 w_i \sigma_i \sigma_m \rho_{i,m} - 2 w_i^2 \sigma_i \sigma_m \rho_{i,m} \right)^{\frac{1}{2}} \\ &= \frac{1}{2} \left(w_i^2 \sigma_i^2 + \sigma_m^2 - 2 w_i \sigma_m^2 + w_i^2 \sigma_m^2 + 2 w_i \sigma_i \sigma_m \rho_{i,m} - 2 w_i^2 \sigma_i \sigma_m \rho_{i,m} \right)^{-\frac{1}{2}} \\ &\quad \frac{\partial}{\partial w_i} \left(w_i^2 \sigma_i^2 + \sigma_m^2 - 2 w_i \sigma_m^2 + w_i^2 \sigma_m^2 + 2 w_i \sigma_i \sigma_m \rho_{i,m} - 2 w_i^2 \sigma_i \sigma_m \rho_{i,m} \right) \\ &= \frac{1}{2\sqrt{\sigma_p^2}} \left(2 w_i \sigma_i^2 - 2 \sigma_m^2 + 2 w_i \sigma_m^2 + 2 \sigma_i \sigma_m \rho_{i,m} - 4 w_i \sigma_i \sigma_m \rho_{i,m} \right) \\ &= \frac{w_i \sigma_i^2 - \sigma_m^2 + w_i \sigma_m^2 + \sigma_i \sigma_m \rho_{i,m} - 2 w_i \sigma_i \sigma_m \rho_{i,m}}{\sigma_p} \\ &= \frac{w_i \left(\sigma_i^2 + \sigma_m^2 \right) - \sigma_m^2 + \sigma_i \sigma_m \rho_{i,m} (1 - 2 w_i)}{\sigma_p}\end{aligned} \tag{158}$$

Let's see what we can learn by plugging $w_i = 0$, the point at which the slope of the investment opportunity curve matches the slope of the CML, and then setting the two slopes equal. Plugging $w_i = 0\%$, Equation 158 simplifies to

$$\frac{\partial \sigma_p}{\partial w_i} = \frac{w_i\left(\sigma_i^2 + \sigma_m^2\right) - \sigma_m^2 + \sigma_i \sigma_m \rho_{i,m}(1 - 2w_i)}{\sigma_p}$$

$$= \frac{\sigma_i \sigma_m \rho_{i,m} - \sigma_m^2}{\sigma_p} \tag{159}$$

$$= \frac{\sigma_i \sigma_m \rho_{i,m} - \sigma_m^2}{\sigma_m}$$

$$= \sigma_i \rho_{i,m} - \sigma_m$$

since $\sigma_p = \sigma_m$ at $w_i = 0$. Using Equations 157 and 159, the slope of the investment opportunity curve at $w_i = 0$ is[15]

$$\frac{\frac{\partial E(R_p)}{\partial w_i}}{\frac{\partial \sigma_p}{\partial w_i}} \Big| (w_i = 0) = \frac{E(R_i) - E(R_m)}{\sigma_i \rho_{i,m} - \sigma_m}$$

Setting the slope of the investment opportunity curve equal to the slope of the CML from Equation 154, we have

$$\frac{E(R_i) - E(R_m)}{\sigma_i \rho_{i,m} - \sigma_m} = \frac{E(R_m) - r_f}{\sigma_m}$$

Rearranging gives us

$$\left[E(R_i) - E(R_m)\right]\sigma_m = \left[E(R_m) - r_f\right]\left(\sigma_i \rho_{i,m} - \sigma_m\right)$$

$$E(R_i)\sigma_m - E(R_m)\sigma_m = E(R_m)\sigma_i \rho_{i,m} - r_f \sigma_i \rho_{i,m} - E(R_m)\sigma_m + r_f \sigma_m$$

$$
\begin{aligned}
E(R_i)\sigma_m &= r_f\sigma_m + \sigma_i\rho_{i,m}\left[E(R_m)-r_f\right] \\
E(R_i) &= r_f + \frac{\sigma_i\rho_{i,m}}{\sigma_m}\left[E(R_m)-r_f\right] \\
&= r_f + \frac{\sigma_i\rho_{i,m}}{\sigma_m}\frac{\sigma_m}{\sigma_m}\left[E(R_m)-r_f\right] \\
&= r_f + \frac{Cov_{i,m}}{\sigma_m^2}\left[E(R_m)-r_f\right] \\
&= r_f + \beta\left[E(R_m)-r_f\right]
\end{aligned} \tag{160}
$$

where $\beta = \frac{\text{cov}_{i,m}}{\sigma_m^2}$. This is the CAPM. It dictates a linear relationship between the expected return on asset i and the *market risk premium* given by $E(R_m)-r_f$. This should be fairly intuitive following what we've learned. We know from the CML that increased risk yields higher expected return, but we've also seen that the idiosyncratic risk of individual assets can be diversified away by including it in a portfolio. The CAPM suggests instead that the expected return of an individual asset increases as its *systemic* risk increases, where β measures systemic risk.

Systemic risk is distinct from idiosyncratic risk. Idiosyncratic risk is specific to a particular company or industry, while systemic risk is endemic to the market as a whole. If $\beta = 1$ for an individual asset, its return should be identical to the market portfolio's. If $\beta = 0$, the individual asset has no linear correlation with the market portfolio and returns the risk-free rate as a result. If $\beta < 0$, the individual asset has negative systemic risk, moving opposite of the market portfolio.

Notice also that the definition of β here matches that of the coefficient $\hat{\beta}$ derived via OLS from Section 6.1! This is convenient, as it facilitates a simple way to test the CAPM theory empirically.[16]

Graphing Equation 160 gives us a fictitious example of the *Security Market Line* (SML).

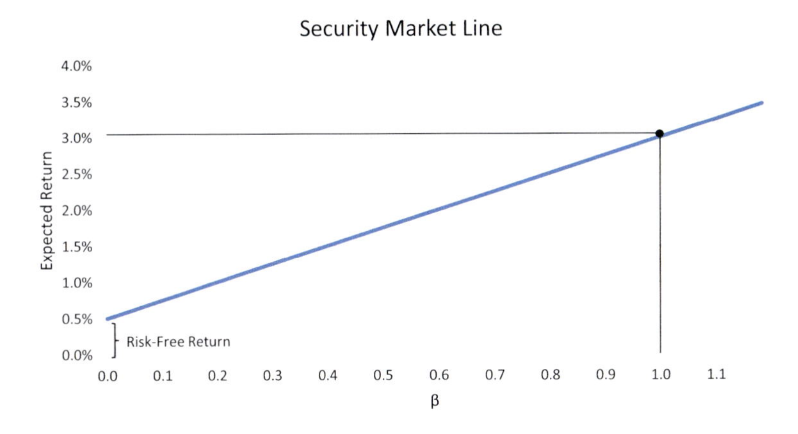

FIGURE 6.12 Security market line.

The SML (the blue line) in Figure 6.12 reflects a risk-free rate $r_f = 0.5\%$ and a return on the market portfolio $R_m = 3.0\%$. The market portfolio is found on the black dot where the black lines intersect. Notice this portfolio has a systemic risk coefficient $\beta = 1$.

All securities that are correctly priced should have an expected return that lies somewhere on the SML, depending on their systemic risk coefficient. Accordingly, one may think of the blue line as the opportunity cost of a particular investment. If a security has an expected return above the SML, this signifies the security is underpriced and gives a buy signal. If a security has an expected return below the SML, this signifies the security is overpriced and gives a sell signal.

Rearranging Equation 160, we have

$$E(R_i) = r_f + \beta\left[E(R_m) - r_f\right]$$

$$E(R_i) - r_f = \beta\left[E(R_m) - r_f\right] \tag{161}$$

$$\frac{E(R_i) - r_f}{\beta} = E(R_m) - r_f$$

The left-hand side of Equation 161 represents the *Treynor ratio*. All securities on the SML have the same Treynor ratio; and since the market portfolio has a systemic risk coefficient $\beta = 1$, the Treynor ratio is equal to the market risk premium, which is the slope of the SML.

NOTES

1 In other words, e_t accounts for the portion of Δy explained by all other unobserved factors.
2 This section is an adaptation of Wooldridge (2018).
3 The Gauss-Markov assumptions cover model specification, sampling approach, and properties of the underlying data. For more on the Gauss-Markov assumptions, see Wooldridge (2018).
4 Here we are finding the value of $\hat{\beta}$ that satisfies the first order condition for optimality. Technically, the first order condition only identifies a stationary point, while the second order condition is required to determine if the stationary point is a maximum or a minimum. We skip the second order condition here.
5 This section is an adaptation of Tsay (2010).
6 One should be careful not to confuse the mean with the mode or median. The mode is the highest point in the distribution (i.e., the most probable value). The median is the center most point of the distribution. With a normal distribution, the mean is equal to the median and the mode.
7 This section is an adaptation of Markowitz (1952) and Sharpe (1964).
8 This definition of portfolio variance relies on a deterministic correlation coefficient ρ. If ρ were stochastic, the portfolio variance would be as well, potentially thickening the tails of the expected return distribution and undermining the normality assumption and use of variance as the appropriate proxy for risk.
9 Since portfolios tend to have dozens of assets or more, optimal weights are often found using linear programming, which is not covered in this book. Another common approach is to capture both objectives (maximize expected return and minimize risk) by choosing security weights to construct a portfolio with maximum Sharpe ratio. The Sharpe ratio is $\frac{E(R_p)-r_f}{\sigma_p}$ where r_f is the risk-free rate. The Sharpe ratio is a dimensionless quantity expressing excess return per unit of risk, and is discussed further in Section 6.2.3.
10 This section is an adaptation of Sharpe (1964) and Tobin (1958).
11 As noted in Concept Refresher 6.1, risk aversion implies that in order to accept more risk, an investor requires more return. As we just learned, the Sharpe ratio quantifies excess return per unit of risk. Being a risk averse investor amounts to preferring the portfolio with the highest Sharpe ratio. Further, if investors care about more than just expected return and standard deviation of return (in favor of, say, a different risk proxy), then they may not have any preference regarding Sharpe ratio.
12 Assigning –50% weight to the risk-free asset signifies “borrowing” at the risk-free rate.

13 As it turns out, the market portfolio represents the portfolio consisting of all investable assets, each with a portfolio weight proportional to its relative market value (i.e., the market value of the asset divided by the sum of market values of all investable assets). In practice, the market portfolio is generally proxied using the S&P 500 or a similar index.

14 This section is an adaptation of Sharpe (1964) and Tobin (1958). For additional context and other avenues to achieve the same end, see Treynor (1961), Lintner (1965a,b), Mossin (1966), and Black (1972).

15 Put simply, the slope of a curve is rise over run. Looking at Figure 6.11, the "rise", or the y-axis, is the expected return, while the "run", or the x-axis, is the standard deviation of return. The slope of the investment opportunity curve is then $\frac{\partial E(R_p)}{\partial \sigma_p}$, which is equal to $\frac{\frac{\partial E(R_p)}{\partial w_i}}{\frac{\partial \sigma_p}{\partial w_i}}$; the numerator and denominator of which was just calculated!

16 Empirical evidence suggests the CAPM does a poor job of explaining actual stock market returns. See Fama and French (2004) for details. This is conceptually attributable to the unrealistic assumptions of the CAPM.

SECTION 7

Uncertainty & Value

Uncertainty and risk are often used interchangeably, though important distinctions have been made between them. In 1921, Frank Knight covered this distinction in detail in his book *Risk, Uncertainty, and Profit*.[1] We introduce two new definitions here, which we will explain using illustrative examples.

Consider a game of Russian roulette; a six-shooter with one bullet in the chamber. You're handed the gun. Can you calculate the probability of life and death? Assuming a working gun and accurate bullet count, you can! Playing this game constitutes a *risk*. You don't know the outcome in advance, but you do know your chances.

Now consider a twist to this scenario. Rather than playing with a gun that's known to be working and carefully loaded, the game is played with a 200-year-old gun dug out of mud. You don't know how many bullets it can hold nor whether there are any bullets in it. Further, you don't know if the gun is even capable of firing anymore. There is still a danger to putting the gun to your head and pulling the trigger, but it is now categorically different than the risk described in the previous scenario. We can no longer calculate the chance of death. We term this *uncertainty* (also known as *Knightian uncertainty*).

Up to this point, we have worked exclusively with risk, not uncertainty. To explore the impact of uncertainty on pricing, let's return to an earlier representation of stochastic return. Recall security S from Section 4.2,

DOI: 10.1201/9781032687650-7

which returns deterministic return r and stochastic return ε_t over time t. The future price is modeled as

$$S_t = S_0 e^{(r+\varepsilon_t)t}$$

In earlier treatment we gave ε_t a particular probabilistic structure, allowing ε_t to be interpreted as return attributable to risk. Now consider ε_t instead as an uncertain return, lacking any structure known in advance. And let's rearrange so that we're solving for present value rather than future.

$$\begin{aligned} S_0 &= \frac{S_t}{e^{(r+\varepsilon_t)t}} \\ &= S_t e^{-(r+\varepsilon_t)t} \end{aligned} \tag{162}$$

What we have in Equation 162 is a stochastic opportunity cost of capital used in discounting. When might this framework be relevant? Primarily when the time horizon is long. The impact of discounting increases as the time horizon increases, meaning small errors in the discount rate can add up to large valuation errors.

In corporate finance, cash flows are generally projected no more than a decade into the future. Private equity funds, for example, tend to have 6–12-year lives, requiring projections in this range for approximation of portfolio company values. In this range, uncertainty over the discount rate is relatively immaterial.

In public finance, projects may be measured in decades. While costs and benefits deep into the future are increasingly uncertain, small errors in cost/benefit assessment are heavily outweighed by the impact of discounting. In other words, one might be extremely uncertain about net benefits 100 years in the future, but the discounting effect is so large over this time that the present value of small or even medium-sized measurement errors might be minuscule. Small errors in the discount rate over this period, on the other hand, can be amplified and have an outsized effect.

To understand why, we first introduce Jenson's Inequality. We then apply what we learn about Jenson's Inequality to a discounting problem.

7.1 JENSON'S INEQUALITY

Jenson's Inequality[2] makes a strong statement about the statistical properties of nonlinear functions, resulting in special treatment of nonlinear functions when dealing with uncertainty.

Consider two functions $g(x)=2x$ and $f(x)=x^2$, $x\in\{1,\dots,10\}$. We visualizing $g(x)=2x$ in Figure 7.1.

Notice the linear nature of this function. If we take the average $\overline{x}$ across its domain, we get

$$\overline{x}=\frac{1}{10}\sum_{x=1}^{10}x=5.5$$

Similarly, taking the average of the function $\overline{g(x)}$ for $x=\{1,\dots,10\}$, we get

$$\overline{g(x)}=\frac{1}{10}\sum_{x=1}^{10}2x=11$$

Notice a convenient symmetry here. To calculate the average of the function $\overline{g(x)}$, we need only calculate the function of the average $g(\overline{x})$:

$$g(\overline{x})=2\overline{x}=2\times 5.5=11=\overline{g(x)}$$

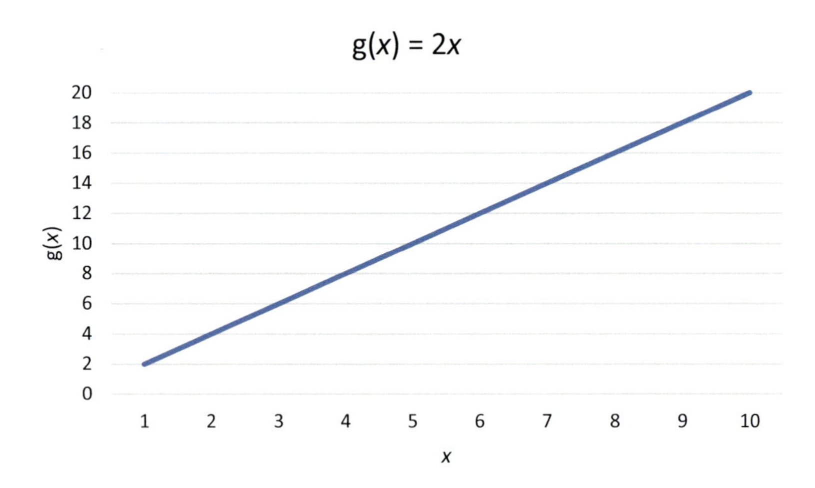

FIGURE 7.1 Visualization of g.

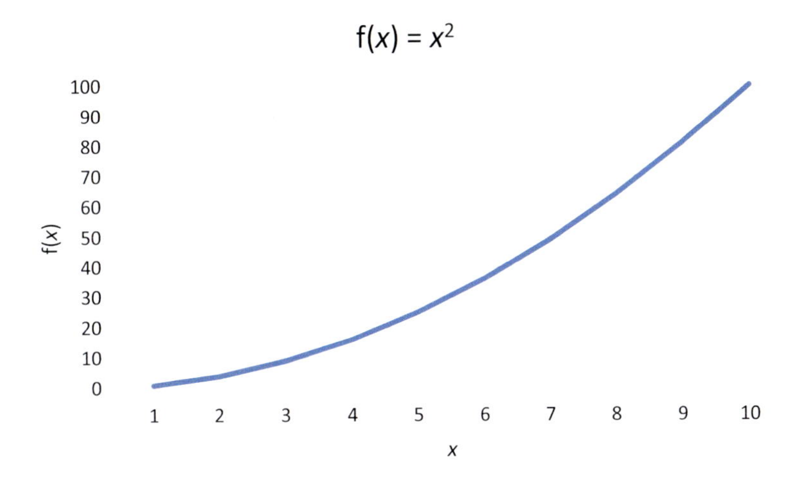

FIGURE 7.2 Visualization of f.

With the linear function $g(x) = 2x$, there is equality between the average of the function and the function of the average (i.e., $\overline{g(x)} = g(\bar{x})$).

Now let's repeat this exercise using the nonlinear function $f(x) = x^2$. We visualizing $f(x) = x^2$ in Figure 7.2.

Notice the nonlinear nature of this function. The average of x across its domain remains $\bar{x} = 5.5$. Taking the average of the function $\overline{f(x)}$ for $x = \{1,\ldots,10\}$, we get

$$\overline{f(x)} = \frac{1}{10}\sum_{x=1}^{10} x^2 = 38.5$$

The symmetry has disappeared. Calculating the function of the average $f(\bar{x})$:

$$f(\bar{x}) = \bar{x}^2 = 5.5^2 = 30.25 < \overline{f(x)}$$

We no longer have equality between the function of the average and the average of the function. Rather $f(\bar{x}) < \overline{f(x)}$. This is an informal demonstration of Jenson's Inequality. Formally, for convex functions, we have[3]

$$f\left[E(x)\right] \leq E\left[f(x)\right]$$

where $E(x) = \bar{x}$ and $E\left[f(x)\right] = \overline{f(x)}$.

CONCEPT REFRESHER 7.1: CONVEX VS. CONCAVE

A function is convex or concave depending on whether a line segment connecting points in the function lies above or below the function on a graph. If the straight line lies above (below) the function, it is convex (concave). The convexity or concavity of a function can be determined mathematically by calculating its second derivative at particular points. A positive second derivative indicates convexity, while a negative second derivative indicates concavity.

This is best demonstrated with an example. Take the function $f(x)=x^2$ over the domain $x \in \{1,\ldots,10\}$ represented by the blue curve in Figure 7.3. The red line segment connecting two points on the curve lies above the line, indicating that $f(x)$ is a convex function. Taking the second derivative, we see that $f''(x)=2$. Since this is positive, it confirms that the function is convex over this domain.

Taking instead the function $f(x)=x^{1/2}$ over the same domain gives us the graph in Figure 7.4. The red line segment now lies below the curve, indicating concavity. The second derivate is now $f''(x)=-\frac{1}{4}x^{-\frac{3}{2}}$. Since this is negative for all values of $x \in \{1,\ldots,10\}$, it confirms that the function is concave over this domain.

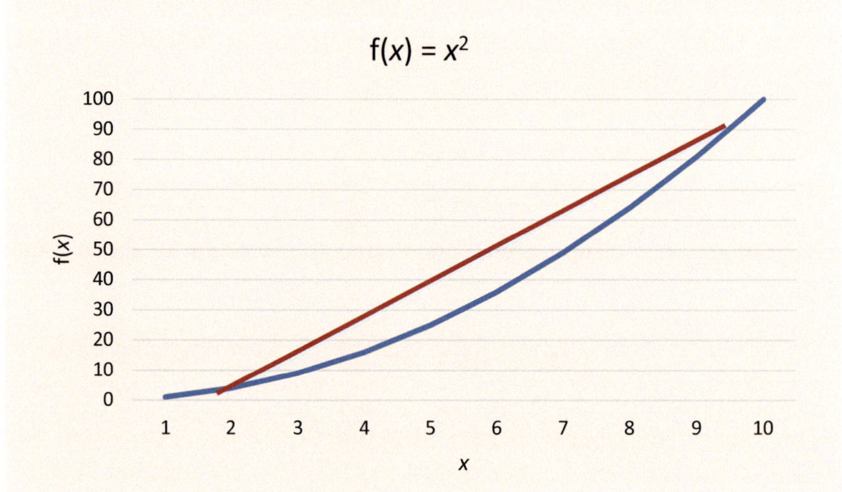

FIGURE 7.3 Visualization of convexity.

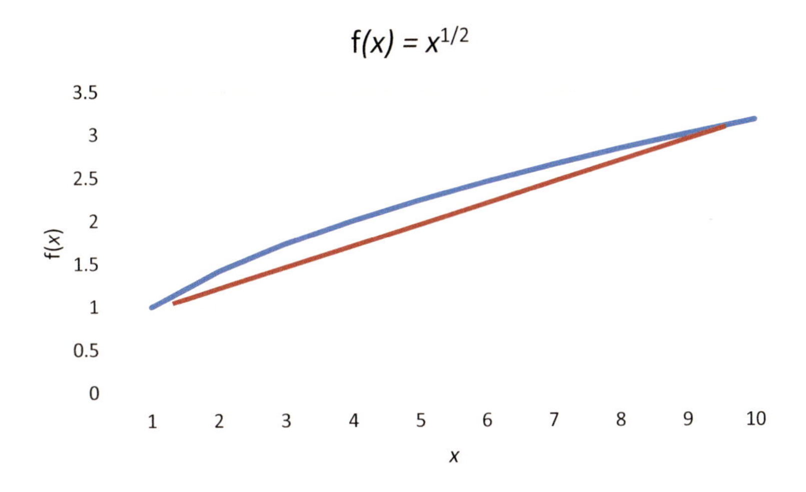

FIGURE 7.4 Visualization of concavity.

7.2 TIME-DECLINING DISCOUNT RATE[4]

Now that we understand the implications of Jenson's Inequality, consider again Equation 162.

$$S_0 = S_t e^{-(r+\varepsilon_t)t}$$

Again, ε_t is now an uncertain rate whose probabilistic structure is unknown. The discounting term $e^{-(r+\varepsilon_t)t}$ is a nonlinear (convex) function. For convenience, we define the variable $\tilde{r} = r + \varepsilon_t$, so that the present value becomes

$$S_0 = S_t e^{-\tilde{r}t}$$

We now want to perform scenario analysis over various finite time horizons to understand the sensitivity of S_0 to changes in $\tilde{r}$. To do this, we consider two time horizons, $t = \{10,100\}$, and three discount rates, $\tilde{r} = \{0.01,0.03,0.05\}$. Since we're dealing with a long time horizon, we set $S_t = 1,000,000$ so that the resulting set of S_0's can be comprehended.

Table 7.1 summarizes the calculated S_0's for each scenario.

We don't know which discount rate $\tilde{r}$ is correct, since by definition $\tilde{r}$ is uncertain, but we've estimated value under three different scenarios. Let's consider these results through the lens of Jenson's Inequality by calculating the average of the discounting function and the function of the average discount rate for both time horizons.

TABLE 7.1 Scenario Analysis (Present Values)

	$\tilde{r}$			
t	**0.01**	**0.03**	**0.05**	**Average**
10	904,837	740,818	606,531	750,729
100	367,879	49,787	6,738	141,468

TABLE 7.2 Scenario Analysis Summary

t	$f(\bar{\tilde{r}})$	$\overline{f(\tilde{r})}$
10	740,818	750,729
100	49,787	141,468

First, we find the function of the average discount rate. Having chosen the set $\{0.01, 0.03, 0.05\}$ for discount rates, the average value of $\tilde{r}$ is $\bar{\tilde{r}} = 0.03$. We know from Table 7.1 that using $\tilde{r} = 0.03$ results in the function values $S_0 = 740{,}818$ for $t = 10$ and $S_0 = 49{,}787$ for $t = 100$.

Next, we find the average of the discounting function. This is already provided in the column labeled Average in Table 7.1 and is found by calculating the equally-weighted average of S_0's for each time horizon. We calculated $\overline{S_0} = 750{,}729$ for $t = 10$ and $\overline{S_0} = 141{,}468$ for $t = 100$.

Table 7.2 restates these results using the notation from Section 7.1.

We see here how much of a difference 90 years makes. At the relatively short time horizon $t = 10$, the convexity of the discounting function $e^{-\tilde{r}t}$ has only a small effect. At the longer time horizon $t = 100$, the convexity effect is much larger. While simply using the average of $\tilde{r}$ for discounting does not cause a large error at relatively short time horizons, it does introduce a rather large error at long time horizons as an estimate for $\overline{f(\tilde{r})}$.

Since we know the function averages $\overline{f(\tilde{r})} = \{750{,}729; 141{,}468\}$ for $t = \{10, 100\}$, respectively, let's back into the value of $\tilde{r}$ necessary to exactly calculate $\overline{f(\tilde{r})}$ for each t. We will term this value the *internal rate of return* $\tilde{r}^*$. In other words, we find $\tilde{r}^*$ in the equations:

$$750{,}729 = 1{,}000{,}000 e^{-10\tilde{r}^*}$$

and

$$141{,}468 = 1{,}000{,}000 e^{-100\tilde{r}^*}$$

With simple algebra, we recover $\tilde{r}^* = \{0.0287, 0.0196\}$ for $t = \{10, 100\}$, respectively. Notice that for $t = 10$, $\tilde{r}^* = 0.0287$ is not so different from the average discount rate $\bar{\tilde{r}} = 0.03$. Conversely, for $t = 100$, $\tilde{r}^*$ is materially lower at 0.0196.

Generalizing, for any set of n uncertain discount rates $\tilde{r} = \{\tilde{r}_1, \ldots, \tilde{r}_n\}$, $\tilde{r}_1 < \tilde{r}_2 < \ldots < \tilde{r}_n$, as the time horizon $t \rightarrow \infty$, the internal rate of return $\tilde{r}^* \rightarrow r_1$. This makes logical sense. As time horizon increases, the present value for scenarios with higher discount rates will accelerate toward zero faster than the present value for scenarios with lower discount rates. This is seen clearly in Table 7.1, where most of the average for $t = 100$ comes from the $\tilde{r} = 0.01$ scenario. This means an increasing portion of the average present value lives in the low discount rate scenarios.

Our conclusion follows. For projects with a sufficiently long time horizon where the opportunity cost of capital is uncertain over the length of the project, a time-declining discount rate should be applied. The thresholds used to separate discount rates might be arbitrary (e.g., r_j for the first 50 years and then r_i for the next 50 years, etc., where $r_i < r_j$). Failure to walk the discount rate down over time will give undue weight to the present value under scenarios with high discount rates because of the convexity of the discounting equation.

NOTES

1 Knight (1948).

2 See Jensen (1906). Proving Jenson's Inequality is beyond the scope of this book.

3 The opposite is true for concave functions (i.e., $f[E(x)] \geq E[f(x)]$).

4 This section is an adaptation of Boardman et al. (2010).

SECTION 8

Capital Structure Irrelevance[1]

We learned in Section 3 how the cost of capital can be used to value firms, projects, and other financial assets. It is natural to ask, "but what should the cost of capital be, and how might our methodology for estimating the cost of capital impact the value of a firm or project?" It turns out this is a difficult question famously answered, to an extent, by Franco Modigliani and Merton Miller in 1958.

To understand the results of Modigliani and Miller, we must first note that every project and every firm can be funded in a number of different ways. We consider two: debt and equity. For our purpose, debt financing represents borrowing at a fixed rate, while equity financing represents the issuance of common stock. The combination of financing used to fund a firm is known as the *capital structure*.

To contrast debt with equity, ponder first the investment risks associated with each. Imagine an investor (providing funds) and an investee (receiving funds). The statutory return to a debt investor is known in advance, seeing as the terms are contractual. Barring default and assuming the investor will not wish to exit the investment early, the investor knows their exact return before committing to the investment. In the event there is a default or bankruptcy, debt investors become the owners of the collateral or assets of the investee.

DOI: 10.1201/9781032687650-8

The return to equity investors, on the other hand, is not known in advance. Should a bankruptcy occur, equity investors have a subordinated claim to the assets of the investee. This means equity investors take losses first.

Given the added safety of debt investments, investors consider debt to be less risky than equity, ceteris paribus. Accordingly, the return one expects from a debt investment in a firm, generally speaking, will be less than that expected from an equity investment in the same firm.

To understand how decisions regarding the capital structure impact the cost of capital, we now define three rates of return:

- Cost of Capital (r_k)
- Cost of Debt (r_d)
- Cost of Equity (r_e)

The cost of capital, already discussed in depth in Section 3, represents the expected rate of profit to a firm as a whole, inclusive of both equity and debt holders. The cost of debt is the expected rate of return on debt investment. The cost of equity is the expected rate of return on equity investment, which should be thought of as the expected rate of profit to a firm as a whole less debt-servicing costs.

To start, let's assume that all firms are 100% equity financed (i.e., no debt). This means $r_k = r_e$. Next, we assume that all firms can be divided into classes. Each class consists of firms whose cost of capital is the same. Said using earlier terminology, the expected rate of profit to each firm in the same class is the same. One might loosely think of these classes as industries, but this need not be an exact analogy.[2] To state this formally for firm i in class j, we have

$$r_{e,j} = \frac{c_i}{E_i} \tag{163}$$

where c_i is the expected profit for firm i and E_i is the value of the equity of firm i. This relationship will hold by definition for all firms in class j, meaning c and E can scale up or down but ultimately are the same proportion of each other. Rearranging Equation 163, we have[3]

$$E_i = \frac{1}{r_{e,j}} c_i \tag{164}$$

Equation 164 makes it explicit that the value of firm i in class j can be found by multiplying its expected profit by the inverse of its cost of capital.

We will now drop the assumption of 100% equity financing and allow for the possibility that firms are financed with both equity and debt. In doing so, we will now denote the value of a firm using V. Specifically, V refers to the *enterprise value*. The enterprise value of a firm represents the combined value of debt and equity. Formally, we can write

$$V_i = E_i + D_i \tag{165}$$

where E_i is the market value of equity of firm i and D_i is the market value of debt of firm i. It would appear that the owners of firm i can impact the composition of enterprise value by choosing the degree to which they fund their business with debt and equity. But what about the enterprise value itself? We already mentioned that the cost of debt is generally lower than the cost of equity. Can owners of a firm find a way to increase enterprise value just by changing the source of financing?

Under certain assumptions about the state of the world, it turns out the answer to this question is *no*. Changing financing cannot increase (or decrease) enterprise value. To see why, we walk through an example. For now, we assume that there are no taxes.

Let's start with two firms in the same class: firm 1 and firm 2. Firm 1's capital structure consists only of equity, while Firm 2's capital structure has some debt. For simplicity, assume these firms have the same expected profit. That is, $c_1 = c_2$. We emphasize that in both cases c represents expected profit *to the firm*, meaning it is available to pay both debt and equity investors (i.e., before interest expense).

The enterprise value of firm 1 is $V_1 = E_1$, consisting entirely of the equity value E_1. The enterprise value of firm 2, $V_2 = E_2 + D_2$, consists of equity value E_2 and debt value D_2. An investor has a portfolio that consists only of s_2 amount of equity in firm 2. Denote $\alpha = \frac{s_2}{E_2}$ the fraction of the total outstanding equity that this investor owns. The return to this portfolio, Y_2, will equal the fraction α of firm 2's expected profit less an interest expense of $r_d D_2$, where r_d is the interest rate paid on debt (i.e., the cost of debt).[4] Since we assumed these firms have the same expected profit, we can denote $c = c_1 = c_2$. Formally, we have

$$Y_2 = \alpha\left(c - r_d D_2\right) \tag{166}$$

Now assume that this investor sold their $s_2 = \alpha E_2$ worth of firm 2 equity and purchased $s_1 = \alpha(E_2 + D_2)$ worth of firm 1. This would be accomplished by using the αE_2 from the sale of firm 2 holdings and borrowing αD_2 on their own account by pledging the new holdings of firm 1 as collateral. The cost of this borrowing would be $\alpha r_d D_2$. This would secure the investor the fraction $\frac{s_1}{E_1} = \frac{\alpha(E_2 + D_2)}{E_1}$ of firm 1's equity and earnings. The return on this newly restructured portfolio, Y_1, is then

$$
\begin{aligned}
Y_1 &= \frac{\alpha(E_2 + D_2)}{E_1} c - \alpha r_d D_2 \\
&= \alpha \frac{V_2}{V_1} c - \alpha r_d D_2 \\
&= \alpha \left(\frac{V_2}{V_1} c - r_d D_2 \right)
\end{aligned} \tag{167}
$$

Contrasting Equation 166 with Equation 167, we see that $Y_1 > Y_2$ when $V_1 < V_2$. In other words, if the enterprise value of firm 2 is greater than that of firm 1, then the return from holding firm 1 in a portfolio would be greater than the return from holding firm 2 in their portfolio. Accordingly, it would be beneficial for an investor in firm 2 to sell their holdings, which would reduce E_2 and by connection V_2, and to purchase a position in firm 1, which would raise E_1 and V_1. This means the levered firm, firm 2, cannot maintain a higher value than the unlevered firm, firm 1, because investors can just achieve the same leverage that firm 2 has by borrowing on their own account as we did above![5]

This is the first major takeaway from Modigliani and Miller. What we have effectively learned is that, in equilibrium, firms in the same class with the same expected return must have the same enterprise value, no matter their capital structure. This means, under the assumptions so far represented, a firm cannot impact its enterprise value by changing its capital structure.

If (by definition) firms in the same class have the same cost of capital r_k, we can write for firm i in class j:

$$
V_i = \frac{c_i}{r_{k,j}} \tag{168}
$$

Note that this follows from Equations 163 and 164, but is now generalized for firms whose capital structures include both equity and debt. Rearranging Equation 168 and using Equation 165, we have

$$r_{k,j} = \frac{c_i}{V_i} = \frac{c_i}{E_i + D_i} \tag{169}$$

We can now restate the cost of equity from Equation 163 in the context of a firm that is not 100% equity financed. Incorporating debt-servicing costs, we have

$$r_{e,j} = \frac{c_i - r_d D_i}{E_i} \tag{170}$$

where $r_{e,j}$ represents the cost of equity for any firm i in class j. We know from Equation 169 that $c_i = r_{k,j}(E_i + D_i)$. Substituting this into Equation 170 gives us

$$\begin{aligned} r_{e,j} &= \frac{c_i - r_d D_i}{E_i} \\ &= \frac{r_{k,j}(E_i + D_i) - r_d D_i}{E_i} \\ &= \frac{r_{k,j} E_i + r_{k,j} D_i - r_d D_i}{E_i} \\ &= r_{k,j} + (r_{k,j} - r_d)\frac{D_i}{E_i} \end{aligned} \tag{171}$$

Equation 171 makes it clear that the cost of equity r_e is a linear function of leverage, given by the debt-to-equity ratio $\frac{D_i}{E_i}$. Said differently, the cost of equity is equal to the cost of capital appropriate for a firm in class j plus a premium related to risk stemming from increased leverage.[6]

Rearranging Equation 171 to isolate the debt-to-equity ratio, we have

$$\frac{D_i}{E_i} = \frac{r_{e,j} - r_{k,j}}{r_{k,j} - r_d} \tag{172}$$

Equipped with Equation 172, let us now find an expression for the cost of capital $r_{k,j}$ that satisfies this equation. Specifically, we will define here the *weighted average cost of capital* (WACC) for firm i in class j to be

$$r_{k,j} = \frac{E_i}{E_i + D_i} r_{e,j} + \frac{D_i}{E_i + D_i} r_d = \frac{r_{e,j} E_i + r_d D_i}{E_i + D_i} \tag{173}$$

Plugging Equation 173 into Equation 172, we have

$$\frac{D_i}{E_i} = \frac{r_{e,j} - r_{k,j}}{r_{k,j} - k_d}$$

$$= \frac{r_{e,j} - \left(\frac{r_{e,j} E_i + r_d D_i}{E_i + D_i} \right)}{\left(\frac{r_{e,j} E_i + r_d D_i}{E_i + D_i} \right) - r_d}$$

$$= \frac{\frac{r_{e,j}(E_i + D_i)}{E_i + D_i} - \left(\frac{r_{e,j} E_i + r_d D_i}{E_i + D_i} \right)}{\left(\frac{r_{e,j} E_i + r_d D_i}{E_i + D_i} \right) - \frac{r_d(E_i + D_i)}{E_i + D_i}}$$

$$= \frac{\frac{r_{e,j}(E_i + D_i) - r_{e,j} E_i - r_d D_i}{E_i + D_i}}{\frac{r_{e,j} E_i + r_d D_i - r_d(E_i + D_i)}{E_i + D_i}}$$

$$= \frac{r_{e,j}(E_i + D_i) - r_{e,j} E_i - r_d D_i}{r_{e,j} E_i + r_d D_i - r_d(E_i + D_i)}$$

$$= \frac{r_{e,j} E_i + r_{e,j} D_i - r_{e,j} E_i - r_d D_i}{r_{e,j} E_i + r_d D_i - r_d E_i - r_d D_i}$$

$$= \frac{r_{e,j} D_i - r_d D_i}{r_{e,j} E_i - r_d E_i} = \frac{D_i (r_{e,j} - r_d)}{E_i (r_{e,j} - r_d)} = \frac{D_i}{E_i}$$

We see here that the WACC from Equation 173 satisfies the identity from Equation 172. This means the cost of capital for a firm should be the weighted average of its cost of equity and cost of debt, where weights are determined by the proportion of the capital structure financed via equity and debt, respectively. Consistent with the results above, however, a firm could not change its WACC by changing its capital structure, as changing the amount of leverage will have an offsetting impact on the cost of equity.

8.1 CAPITAL BUDGETING

Tying the results from Section 8 back to the results from Section 3.1, we can say that a firm in class j should pursue a project if and only if the rate of return on the project r^* is greater than or equal to the cost of capital $r_{k,j}$, estimated using the WACC. This is why the cost of capital is sometimes referred to as the *hurdle rate*. And as it turns out, it doesn't matter how the project is financed for this hurdle rate to remain valid. To see that this is true, denote the enterprise value V_0 of a firm in class j at time $t=0$ and restate Equation 168 to be

$$V_0 = \frac{c_0}{r_{k,j}} \tag{174}$$

where c_0 is the expected profit at time $t=0$ and $r_{k,j}$ is the cost of capital for a firm in class j. The value of equity at time $t=0$ is then

$$E_0 = V_0 - D_0 \tag{175}$$

where again D_0 is the value of debt at time $t=0$. The firm is now assumed to borrow funds for a project costing B dollars with rate of return r^*. This means the firm's expected profits should grow by Br^*. Using Equation 174, the firm's enterprise value at time $t=1$ is then

$$\begin{aligned} V_1 &= \frac{c_0 + Br^*}{r_{k,j}} = \frac{c_0}{r_{k,j}} + \frac{Br^*}{r_{k,j}} \\ &= V_0 + \frac{Br^*}{r_{k,j}} \end{aligned} \tag{176}$$

By borrowing B, the total debt has increased by B. Consequently, using Equations 175 and 176, the value of equity at time $t=1$ can be written

$$\begin{aligned}
E_1 &= V_1 - (D_0 + B) \\
&= V_0 + \frac{Br^*}{r_{k,j}} - D_0 - B \\
&= V_0 - D_0 + \frac{Br^*}{r_{k,j}} - B \\
&= E_0 + \frac{Br^*}{r_{k,j}} - B \\
&= E_0 + B\left(\frac{r^*}{r_{k,j}} - 1\right)
\end{aligned} \tag{177}$$

It is clear from Equation 177 that the project will increase the value of equity when $r^* > r_{k,j}$. Otherwise, the project will decrease the value of equity. Now take the same scenario but assume that the project is funded entirely through equity issuance. Let P_0 denote the market price per share of common stock at time $t = 0$, reflecting expectations at this time, and let N be the original number of shares. Then

$$P_0 = \frac{E_0}{N} \tag{178}$$

where E_0 is again the equity value at time $t = 0$. The number of new shares M needed to finance a project costing B dollars is given by

$$M = \frac{B}{P_0} \tag{179}$$

Using Equations 174, 175, and 178, the market value of equity at time $t = 1$ can be expressed as

$$
\begin{aligned}
E_1 &= \frac{c_0 + r^* B}{r_{k,j}} - D_0 \\
&= \frac{c_0}{r_{k,j}} + \frac{r^* B}{r_{k,j}} - D_0 \\
&= V_0 + \frac{r^* B}{r_{k,j}} - D_0 \\
&= V_0 - D_0 + \frac{r^* B}{r_{k,j}} \\
&= E_0 + \frac{r^* B}{r_{k,j}} \\
&= NP_0 + \frac{r^* B}{r_{k,j}}
\end{aligned} \tag{180}
$$

Since we know from Equation 179 that $B = MP_0$, and using Equation 180, the price per share at time $t = 1$ becomes

$$
\begin{aligned}
P_1 &= \frac{E_1}{N+M} \\
&= \frac{NP_0 + \frac{r^* B}{r_{k,j}}}{N+M} \\
&= \frac{NP_0 + \frac{r^* B}{r_{k,j}} - B + MP_0}{N+M} \\
&= \frac{(N+M)P_0 + \left(\frac{r^*}{r_{k,j}} - 1 \right) B}{N+M} \\
&= P_0 + \frac{\left(\frac{r^*}{r_{k,j}} - 1 \right) B}{N+M}
\end{aligned} \tag{181}
$$

We can see from Equation 181 that $P_1 > P_0$ when $r^* > r_{k,j}$. This proves that $r_{k,j}$ is the appropriate hurdle rate whether a project is funded via equity issuance or debt.

To this point, we have operated under the assumption that there are no taxes. Introducing taxes and the tax-deductibility of interest, we denote the tax rate τ and define the after-tax expected profit c_i^τ to firm i

$$c_i^\tau = (c_i - r_d D_i)(1-\tau) + r_d D_i \tag{182}$$

We define also a new cost of capital r_k^τ using this after-tax profit

$$r_k^\tau = \frac{c_i^\tau}{V_i} \tag{183}$$

Restating Equation 182 to solve for the pre-tax expected profit c_i for firm i, we have

$$c_i = \frac{c_i^\tau - r_d D_i}{(1-\tau)} + r_d D_i \tag{184}$$

Using Equations 183 and 184, we can restate the cost of capital r_k from Equation 169 for firm i in class j as[7]

$$\begin{aligned}
r_{k,j} &= \frac{c_i}{V_i} \\
&= \frac{c_i^\tau - r_d D_i}{V_i(1-\tau)} + \frac{r_d D_i}{V_i} \\
&= \frac{c_i^\tau}{V_i(1-\tau)} - \frac{r_d \frac{D_i}{V_i}}{(1-\tau)} + r_d \frac{D_i}{V_i} \\
&= \frac{r_{k,j}^\tau}{(1-\tau)} - \frac{r_d \frac{D_i}{V_i}}{(1-\tau)} + r_d \frac{D_i}{V_i} \frac{(1-\tau)}{(1-\tau)} \\
&= \frac{r_{k,j}^\tau - r_d \frac{D_i}{V_i} + r_d \frac{D_i}{V_i}(1-\tau)}{(1-\tau)} \\
&= \frac{r_{k,j}^\tau + r_d \frac{D_i}{V_i}((1-\tau)-1)}{(1-\tau)} \\
&= \frac{r_{k,j}^\tau - r_d \frac{D_i}{V_i}\tau}{(1-\tau)}
\end{aligned} \tag{185}$$

Equation 185 shows that the cost of capital now depends on the debt capitalization D_i/V_i. As the debt capitalization increases, the cost of capital decreases. This contradicts our earlier results! When interest on debt becomes tax deductible, and the tax rate is not zero, firm value can increase from adding more debt in the capital structure. Now we can repeat our exercise to determine how different funding sources impact the cost of capital when interest on debt is tax deductible.

If a new project were financed entirely by new common stock, we must simply plug $D_i = 0$ into Equation 185 to get

$$r_{k,j}^{E} = \frac{r_{k,j}^{\tau} - r_d \frac{0}{V_i}\tau}{(1-\tau)} = \frac{r_{k,j}^{\tau}}{(1-\tau)} \tag{186}$$

where the superscript E indicates that this is the cost of capital under 100% equity financing. If instead, we assume 100% debt financing, then $D_i = V_i$ and using Equation 186 the cost of capital becomes

$$\begin{aligned} r_{k,j}^{D} &= \frac{r_{k,j}^{\tau} - r_d \frac{V_i}{V_i}\tau}{(1-\tau)} = \frac{r_{k,j}^{\tau} - r_d\tau}{(1-\tau)} \\ &= \frac{r_{k,j}^{\tau}}{(1-\tau)} - \frac{r_d\tau}{(1-\tau)} \\ &= r_{k,j}^{E} - \frac{r_d\tau}{(1-\tau)} \end{aligned} \tag{187}$$

Notice from Equations 186 and 187 that $r_{k,j}^{D} < r_{k,j}^{E}$ when $0 < \tau \le 1$. This means the cost of capital for a debt-funded project is lower than the cost of capital for an equity-funded project when taxes are non-zero and the interest on debt is tax deductible! We should also take note of the paradoxical conclusion that the higher the cost of debt r_d, the lower the cost of capital $r_{k,j}^{D}$ for a debt-funded project.

One final note on this topic. Assuming a non-zero, positive tax rate, it is noteworthy to contrast the enterprise value of a firm with and without debt in its capital structure. Denote the enterprise value of an unlevered firm V_U and the enterprise value of a levered firm V_L. We will still use c for expected profit, r_k for the cost of capital, r_d for the cost of debt, D for the value of debt, and τ for the tax rate. Then, for an unlevered firm in class j, we have

$$V_U = \frac{c(1-\tau)}{r_{k,j}} \tag{188}$$

and using Equation 188, for a levered firm in class j we have

$$\begin{aligned} V_L &= \frac{(c - r_d D)(1-\tau) + r_d D}{r_{k,j}} \\ &= \frac{c - r_d D - c\tau + r_d D\tau + r_d D}{r_{k,j}} \\ &= \frac{c - c\tau + r_d D\tau}{r_{k,j}} \\ &= \frac{c(1-\tau) + r_d D\tau}{r_{k,j}} \\ &= \frac{c(1-\tau)}{r_{k,j}} + \frac{r_d D\tau}{r_{k,j}} \\ &= V_U + \frac{r_d D\tau}{r_{k,j}} = V_U + V_\tau \end{aligned}$$

where $V_\tau = \frac{r_d D\tau}{r_{k,j}}$ is referred to as the value of the *tax shield.*

NOTES

1 Much of this section is adapted from Modigliani and Miller (1958).

2 Categorizing firms based on cost of capital theoretically groups firms by risk. Firms in the same class can be thought of as equally risky from the standpoint of delivering expected profit.

3 Notice the similarities (and differences) between Equation 164 and Equations 17 and 22 from Section 3.2. One might think Modigliani and Miller are assuming a growth rate of zero, but this is not the case. Rather, they defined c_i in a special way that curtails the need to consider growth. If firm i's operations generate profits $c_i(t)$ at times $t = 1, \ldots, T$, then $c_i = \lim_{T\to\infty} \frac{1}{T} \sum_{t=1}^{T} c_i(t)$.

That is, c_i actually represents the time-average profit until the end of firm i's existence. Accordingly, $r_{e,j}$ is literally interpreted as the average cost of equity (or in this case the average cost of capital, as firms here are assumed 100% equity financed such that $r_e = r_k$).

4 Note that r_d is a constant and not indexed to a particular firm. This is because of embedded assumptions regarding debt markets. Specifically, we assume that bond yields are all constant with no risk of loss, and bonds are traded in perfect markets. This latter assumption implies that the "law of one price" holds true. That is, bonds with the same fixed payment trade at the exact same price. The combination of these assumptions mean bonds must all yield the same rate of return, which we denote here as r_d.

5 A similar exercise can be performed when assuming $V_1 > V_2$, and the same result will be found.

6 Increased leverage can be thought of as raising the risk of financial distress, leading equity owners to demand a higher rate of return for taking on this risk. Note that if we relax the assumptions from Endnote 4, then we allow for the possibility that increased leverage could raise the cost of debt k_d as well. In this case, the cost of equity from Equation 171 would no longer be a strictly linear function of firm leverage. Rather, increases in r_d would paradoxically counteract increases in leverage, making the impact of rising leverage uncertain.

7 We solve for r_k because the appropriate cost of capital for investment decision-making should still use the before-tax expected profit.

SECTION 9

Probability of Default

As eluded to in Endnote 5 from Section 3, it can be useful to consider the opportunity cost of capital r in two components: a base rate and a spread. As we saw in Section 3.1, the opportunity cost of capital should reflect the return required by the market to compensate for perceived risk.

To accommodate this view explicitly, we might say that the opportunity cost of capital consists of a base (risk-free) rate r_f and a credit spread s. The risk-free rate r_f is the rate one expects on an investment that bears no risk of loss. The credit spread provides an increased return commensurate with increased risk. We have

$$r = r_f + s$$

Let's see what we can learn by re-framing the present value function after considering the risk of loss. We start by valuing a security with a single future cash flow C based on a continuous compounding rate over a time period t. The present value P_0 of such a security, assuming no risk of loss, is

$$P_0 = Ce^{-r_f t} \tag{189}$$

Note that we are using r_f, the risk-free rate, as our opportunity cost of capital since there is no risk of loss. Now say there is some probability p that the party responsible for paying C will default on its payment. The present value P_0' of this risky security can be written in two ways:

$$P_0' = Ce^{-(r_f + s)t} \tag{190}$$

DOI: 10.1201/9781032687650-9

$$P_0' = (1-p)P_0 + p \times 0 \tag{191}$$

Equation 190 incorporates this risk of loss through the credit spread, while Equation 191 incorporates risk of loss probabilistically. For now, Equation 191 assumes that in the case of default, the value received would be zero. Said differently, the recovery rate is assumed to be zero. Substituting P_0 from Equation 189 into Equation 191, we get

$$\begin{aligned} P_0' &= (1-p)P_0 + p \times 0 \\ &= (1-p)Ce^{-r_f t} \end{aligned} \tag{192}$$

Setting Equation 190 equal to Equation 192 gives us

$$Ce^{-(r_f+s)t} = (1-p)Ce^{-r_f t}$$

Let's solve for the probability of default p.

$$e^{-(r_f+s)t} = (1-p)e^{-r_f t}$$

$$e^{-r_f t}e^{-st} = (1-p)e^{-r_f t}$$

$$e^{-st} = 1-p$$

$$p = 1-e^{-st}$$

We have successfully stated the probability of default in terms of just the credit spread for cases where default implies a complete loss of value. In the real world, investors facing a defaulting party will generally recover some portion of what they're owed. If the recovery rate is denoted b, we can re-state the risky present value from Equations 190 and 191 to be

$$P_0' = Ce^{-(r_f+s)t} \tag{193}$$

$$P_0' = (1-p)P_0 + pbP_0 \tag{194}$$

Once again inserting Equation 189, we expand Equation 194

$$\begin{aligned} P_0' &= (1-p)P_0 + pbP_0 \\ &= (1-p)Ce^{-r_f t} + pbCe^{-r_f t} \\ &= Ce^{-r_f t} - pCe^{-r_f t} + pbCe^{-r_f t} \end{aligned} \tag{195}$$

Setting Equation 193 equal to Equation 195, we have

$$\begin{aligned} Ce^{-(r_f+s)t} &= Ce^{-r_f t} - pCe^{-r_f t} + pbCe^{-r_f t} \\ e^{-r_f t - st} &= e^{-r_f t} - pe^{-r_f t} + pbe^{-r_f t} \\ e^{-r_f t}e^{-st} &= e^{-r_f t} - pe^{-r_f t} + pbe^{-r_f t} \\ e^{-st} &= 1 - p + pb \\ p - pb &= 1 - e^{-st} \\ p(1-b) &= 1 - e^{-st} \\ p &= \frac{1-e^{-st}}{1-b} \end{aligned}$$

The probability of default is now stated in terms of the credit spread and recovery rate.[1] This is convenient, as bond markets are highly liquid[2] and provide most of the information needed to solve for default probabilities. Prices and cash flows (for fixed-rate bonds) are observable, and the proxies commonly chosen to represent the risk-free rate are also observable. While recovery rates are not directly observable, investors can use historical recovery rates or idiosyncratic collateral information to estimate this. Common values for recovery rates are between $b = 0.4$ and $b = 0.6$.

9.1 HAZARD RATES[3]

The probability of default calculated in Section 9 is specific to the time period t. Rather, one may be interested in the probability that default occurs within a particular window of time. That is, consider discrete times $s_1, \ldots, s_n, t$. If we are interested in the probability of default occurring, for example, between times s_1 and s_2 (see Figure 9.1), we use the term *hazard rate* or *failure rate* for this period.

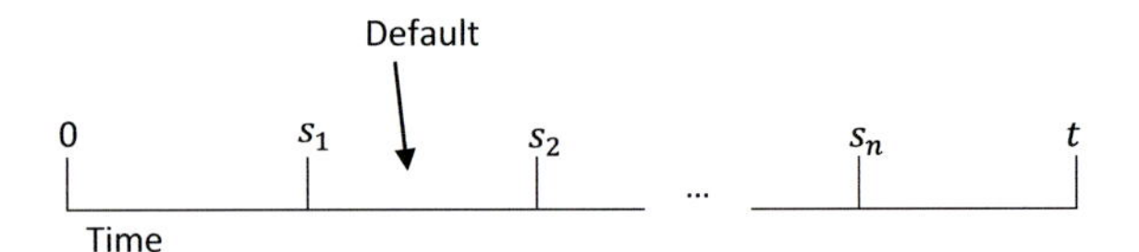

FIGURE 9.1 Hazard rate visualization.

To treat hazard rates more formally, we must envision that the time between s_1 and s_2 gets progressively smaller, even infinitely smaller. Let's start from scratch to understand hazard rates more rigorously. Let X denote the event that default occurs. Let t and h index time. Then the hazard rate $r(t)$ is defined as

$$r(t) = \lim_{h \to 0} \frac{P\big(X \in (t, t+h) \mid X > t\big)}{h} \tag{196}$$

Breaking down Equation 196, X is a random variable denoting a default event. The term $X \in (t, t+h)$ specifies that default occurs in the period between time t and time $t+h$. Thus, $P\big(X \in (t, t+h) \mid X > t\big)$ gives the probability that default occurs in the period between time t and time $t+h$ conditional on default not occurring prior to time t. Dividing by the time increment h scales this probability to a unit amount of time. Then the hazard rate $r(t)$ is the limit of this scaled probability as h gets infinitely small.

Using Bayes' Theorem[4] and some algebra, we have

$$\begin{aligned} r(t) &= \lim_{h \to 0} \frac{P\big(X \in (t, t+h) \mid X > t\big)}{h} \\ &= \lim_{h \to 0} \frac{P\big(X \in (t, t+h), X > t\big)}{hP(X > t)} \\ &= \lim_{h \to 0} \frac{P\big(X \in (t, t+h)\big)}{hP(X > t)} \\ &= \lim_{h \to 0} \frac{F(t+h) - F(t)}{h\big(1 - F(t)\big)} \\ &= \frac{1}{1 - F(t)} \lim_{h \to 0} \frac{F(t+h) - F(t)}{h} \\ &= \frac{f(t)}{1 - F(t)} \end{aligned} \tag{197}$$

Note here that the event $\{X > t\}$ is superfluous in the joint event $\{X \in (t, t+h), X > t\}$, allowing it to be reduced to just $\{X \in (t, t+h)\}$. Note also that $F(\cdot)$ and $f(\cdot)$ are the distribution function and density function, respectively, of X. Specifically, $F(t)$ is the probability that default has occurred prior to time t, while $f(t)$ can roughly be interpreted as the probability of default occurring at time t. This means $1 - F(t)$ is the probability that default has *not* occurred by time t.

We conclude from Equation 197 that the hazard rate is the ratio of the probability of default at time t to the probability of surviving to time t without default. Said differently, $r(t)$ denotes the conditional probability that an entity of age t will default.

It turns out we can learn even more about hazard rates. First, write Equation 197 as[5]

$$r(t) = \frac{f(t)}{1 - F(t)} = \frac{\frac{d}{dt}F(t)}{1 - F(t)}$$

Integrating both sides, we have

$$\int_0^t r(t)\, dt = \int_0^t \frac{\frac{d}{dt}F(t)}{1 - F(t)}\, dt \tag{198}$$

To solve this integral, notice that

$$\frac{d}{dt}\ln\left(1 - F(t)\right) = \frac{1}{1 - F(t)}\left(-F'(t)\right) + k = -\frac{\frac{d}{dt}F(t)}{1 - F(t)} + k \tag{199}$$

by the chain rule, where k is a constant. The right-hand side of Equation 199 is the negative of the integration term in the right-hand side of Equation 198. Hence, we see that

$$\int_0^t r(t)\, dt = \int_0^t \frac{\frac{d}{dt}F(t)}{1 - F(t)}\, dt = -\ln\left(1 - F(t)\right) + k$$

$$-\int_0^t r(t)\, dt + k = \ln\left(1 - F(t)\right)$$

where k is still added since it is an arbitrary constant. Continuing to simplify, we have

$$\exp\left(-\int_0^t r(t)dt + k\right) = 1 - F(t)$$

$$1 - F(t) = e^k \exp\left(-\int_0^t r(t)dt\right)$$

$$F(t) = 1 - e^k \exp\left(-\int_0^t r(t)dt\right)$$

Since we know $F(0) = 0$, letting $t = 0$ gives us

$$F(0) = 1 - e^k \exp\left(-\int_0^0 r(t)dt\right) = 0$$

$$1 = e^k \exp\left(-\int_0^0 r(t)dt\right)$$

$$1 = e^k e^0 = e^k$$

which shows that $k = 0$, meaning

$$F(t) = 1 - \exp\left(-\int_0^t r(t)dt\right) \tag{200}$$

In the special case where the hazard rate is assumed to be a constant c, Equation 200 becomes

$$F(t) = 1 - \exp\left(-c\int_0^t dt\right) = 1 - e^{-ct}$$

which means X is exponential with rate c. This is an interesting case because of the memoryless property[6] of the exponential distribution, which says that the remaining time to default is the same no matter how much time has already passed.

NOTES

1 For relatively small spreads, $p \approx \frac{s}{1-b}$. As s increases this approximation becomes less accurate.
2 Credit spreads might also incorporate an illiquidity premium for securities that are not traded frequently.
3 This section is an adaptation of Ross (2019).
4 Bayes' Theorem defines a relationship between conditional and joint probability. For two events A and B, where $P(A|B)$ denotes the probability of event A conditional on event B and $P(A,B)$ the joint probability of events A and B, Bayes' Theorem states

$$P(A|B)P(B) = P(B|A)P(A) = P(A,B)$$

meaning, for example, that

$$P(A|B) = \frac{P(A,B)}{P(B)}$$

5 Recall in calculus that, for a continuous random variable, the density function is equal to the derivative of the distribution function. For a refresher on these concepts, see Concept Refreshers 4.1 and 4.2.
6 A random variable X is *memoryless* if, for arbitrary time increments s and t, we have

$$P(X > t + s \mid X > t) = P(X > s)$$

It is as if the variable X doesn't "remember" that t time has passed. Note also by Bayes' Theorem that

$$P(X > t + s \mid X > t) = \frac{P(X > t + s, X > t)}{P(X > t)}$$

Setting these two expressions for $P(X > t + s \mid X > t)$ equal gives us

$$\frac{P(X > t + s, X > t)}{P(X > t)} = P(X > s)$$

$$P(X > t + s, X > t) = P(X > s)P(X > t)$$

While the proof is beyond the scope of this text, it turns out this identity is only satisfied when X is exponentially distributed. That is,

$$e^{-c(s+t)} = e^{-cs}e^{-ct}$$

is the only solution, where c is the rate. This means the exponential distribution is the only distribution possessing the memoryless property!

SECTION 10

Appendix

WE END THIS BOOK with a series of miscellaneous topics that didn't fit nicely into an earlier section.

10.1 RULE OF 72

The Rule of 72 is meant to be a quick heuristic approximating how long it takes to double an investment, given a fixed annual return. Formally, it says

$$\frac{72}{r} \approx n$$

where r is the annually compounded interest rate expressed in percentage points (i.e., $4\% = 4$) and n is the number of years until your investment doubles in value. Writing this relation differently, we have

$$rn \approx 72$$

Given a starting investment P, and knowing that n is the number of years it takes to double the value of your investment, we know that

$$P\left(1 + \frac{r}{100}\right)^n = 2P$$

DOI: 10.1201/9781032687650-10

TABLE 10.1 Interest Rates r Necessary to Double Investment for Horizons n

n	$f(n)$	r (%)	n	$f(n)$	r (%)	n	$f(n)$	r (%)
1	100.0	100.0	11	71.5	6.5	21	70.5	3.4
2	82.8	41.4	12	71.4	5.9	22	70.4	3.2
3	78.0	26.0	13	71.2	5.5	23	70.4	3.1
4	75.7	18.9	14	71.1	5.1	24	70.3	2.9
5	74.3	14.9	15	70.9	4.7	25	70.3	2.8
6	73.5	12.2	16	70.8	4.4	26	70.2	2.7
7	72.9	10.4	17	70.7	4.2	27	70.2	2.6
8	72.4	9.1	18	70.7	3.9	28	70.2	2.5
9	72.1	8.0	19	70.6	3.7	29	70.1	2.4
10	71.8	7.2	20	70.5	3.5	30	70.1	2.3

Solving for r gives us

$$P\left(1+\frac{r}{100}\right)^n = 2P$$

$$\left(1+\frac{r}{100}\right)^n = 2$$

$$1+\frac{r}{100} = 2^{\frac{1}{n}}$$

$$r = 100\left(2^{\frac{1}{n}} - 1\right)$$

Multiplying by n, we get

$$rn = 100n\left(2^{\frac{1}{n}} - 1\right) \qquad (201)$$

The question now is, does $100n\left(2^{\frac{1}{n}} - 1\right) \approx 72$? Let's plug different values of n into Equation 201 to see how close this approximation gets us. Labeling $100n\left(2^{\frac{1}{n}} - 1\right) = f(n)$, Table 10.1 lists $f(n)$ and $r = \frac{f(n)}{n}$ for each value of n.

Reviewing $f(n)$ in Table 10.1, using standard rounding convention, we see that 72 is most accurate when the interest rate r ranges between 6.5%

TABLE 10.2 Rules of Thumb

Rate Range (%)	Rule of __
0.1–0.4	69
0.4–3.4	70
3.4–6.5	71
6.5–9.4	72
9.4 12.2	73
12.2–15.2	74
>15.2	Don't use

and 9% (which corresponds to n in the range of 8–11 years). When the interest rate is between 3.4% and 6.5%, the rule of 71 would be most accurate. For an interest rate below 3.4%, the rule of 70 would be most accurate. Table 10.2 dictates what rule should be used for different ranges of annually compounding interest rates.

If the interest rate is continuously compounding, we have

$$Pe^{rn} = 2P$$

$$e^{rn} = 2$$

$$rn = \ln(2)$$

$$rn \approx 69.3$$

The rule of 69.3 is most accurate for a continuously compounding rate, no matter what range r is in.

10.2 QUADRATIC EQUATION

The quadratic equation is useful for finding roots of a polynomial function of degree two (i.e., a quadratic function).[1] These functions appear frequently in finance, the roots of which reveal certain properties of the function. The general form of a quadratic function is

$$ax^2 + bx + c$$

To recover the well-known quadratic equation, we set the function equal to zero and perform algebra to solve for x. We start by dividing through by a:

$$x^2 + \frac{b}{a}x + \frac{c}{a} = 0$$

Adding and subtracting $\left(\frac{b}{2a}\right)^2$ and then rearranging, we get

$$\begin{aligned} x^2 + \frac{b}{a}x + \frac{c}{a} + \left(\frac{b}{2a}\right)^2 - \left(\frac{b}{2a}\right)^2 &= 0 \\ x^2 + \left(\frac{b}{2a}\right)^2 + \frac{b}{a}x + \frac{c}{a} - \left(\frac{b}{2a}\right)^2 &= 0 \end{aligned} \tag{202}$$

We can simplify the blue term in Equation 202 to $\left(x + \frac{b}{2a}\right)^2$ since

$$\begin{aligned} \left(x + \frac{b}{2a}\right)^2 &= \left(x + \frac{b}{2a}\right)\left(x + \frac{b}{2a}\right) \\ &= x^2 + x\frac{b}{2a} + x\frac{b}{2a} + \left(\frac{b}{2a}\right)\left(\frac{b}{2a}\right) \\ &= x^2 + \frac{2xb}{2a} + \left(\frac{b}{2a}\right)^2 \\ &= x^2 + \frac{b}{a}x + \left(\frac{b}{2a}\right)^2 \end{aligned}$$

Making this substitution, we get

$$x^2 + \left(\frac{b}{2a}\right)^2 + \frac{b}{a}x + \frac{c}{a} - \left(\frac{b}{2a}\right)^2 = 0$$

$$\left(x + \frac{b}{2a}\right)^2 + \frac{c}{a} - \frac{b^2}{4a^2} = 0$$

Multiplying c/a by $4a/4a$ allows us to equate the denominators between the last two terms.

$$\left(x+\frac{b}{2a}\right)^2+\frac{c}{a}\frac{4a}{4a}-\frac{b^2}{4a^2}=0$$

$$\left(x+\frac{b}{2a}\right)^2+\frac{4ac-b^2}{4a^2}=0$$

$$\left(x+\frac{b}{2a}\right)^2=-\frac{4ac-b^2}{4a^2}$$

$$\left(x+\frac{b}{2a}\right)^2=\frac{b^2-4ac}{4a^2}$$

$$x+\frac{b}{2a}=\sqrt{\frac{b^2-4ac}{4a^2}}$$

$$x=-\frac{b}{2a}\pm\sqrt{\frac{b^2-4ac}{4a^2}}$$

$$x=-\frac{b}{2a}\pm\frac{\sqrt{b^2-4ac}}{2a}$$

$$x=\frac{-b\pm\sqrt{b^2-4ac}}{2a}$$

10.3 FORWARD RATES FROM SPOT RATES

Consider a term structure of spot rates.[2] For example, say the continuously compounded one- and two-year spot rates are as follows:

Term (year)	Rate (%)
1	3.0
2	3.5

If we were to invest \$100 in an asset earning the one-year spot rate, using what we learned in Section 2.2, at the end of one year we would have

$$100e^{(0.03)1}=103.05$$

Similarly, if we invest \$100 in an asset earning the two-year spot rate, at the end of two years we would have

$$100e^{(0.035)2}=107.25$$

Theoretically, one should be able to achieve the two-year spot rate by reinvesting the 103.05 earned at the end of the first year into an asset earning the *one-year spot rate in one year.* Modeling this, we have

$$103.05e^{r_F} = 107.25$$

$$e^{r_F} = \frac{107.25}{103.05}$$

$$r_F = \ln\left(\frac{107.25}{103.05}\right) = 0.04$$

r_F here is referred to as the *forward rate*. In this example, it represents the return one implicitly expects to earn for a one-year investment that starts in one year. Writing the example differently, we have[3]

$$100 \times \left(e^{0.03}\right)\left(e^{r_F}\right) = 100 \times \left(e^{0.035 \times 2}\right)$$

Generalizing for any two time horizons T_1 and T_2, $T_1 < T_2$, with spot rates r_1 and r_2, respectively, we have

$$e^{r_1 T_1} e^{r_F (T_2 - T_1)} = e^{r_2 T_2}$$

Again, r_F reflects the implied forward rate spanning the time period starting at T_1 and ending at T_2. Solving for r_F, we get

$$e^{r_F (T_2 - T_1)} = \frac{e^{r_2 T_2}}{e^{r_1 T_1}}$$

$$r_F \left(T_2 - T_1\right) = \ln\left(\frac{e^{r_2 T_2}}{e^{r_1 T_1}}\right)$$

$$= \ln\left(e^{r_2 T_2}\right) - \ln\left(e^{r_1 T_1}\right)$$

$$= r_2 T_2 - r_1 T_1$$

meaning

$$r_F = \frac{r_2 T_2 - r_1 T_1}{T_2 - T_1}$$

This allows the implied continuously compounded forward rate to be easily derived from continuously compounded spot rates. To determine the relative position of forward rates and spot rates, we rearrange them as follows:

$$
\begin{aligned}
r_F &= \frac{r_2T_2 - r_1T_1}{T_2 - T_1} \\
&= \frac{r_2T_2}{T_2 - T_1} - \frac{r_1T_1}{T_2 - T_1} \\
&= \frac{r_2T_2 - r_2T_1 + r_2T_1}{T_2 - T_1} - \frac{r_1T_1}{T_2 - T_1} \\
&= \frac{r_2(T_2 - T_1) + r_2T_1}{T_2 - T_1} - \frac{r_1T_1}{T_2 - T_1} \\
&= \frac{r_2(T_2 - T_1)}{T_2 - T_1} + \frac{r_2T_1}{T_2 - T_1} - \frac{r_1T_1}{T_2 - T_1} \\
&= r_2 + \frac{r_2T_1 - r_1T_1}{T_2 - T_1} \\
&= r_2 + \frac{(r_2 - r_1)T_1}{T_2 - T_1}
\end{aligned}
$$

In this form, we can see that the forward curve will lie above the spot curve when $r_2 > r_1$, lie below the spot curve when $r_2 < r_1$, and equal the spot curve when $r_2 = r_1$.

10.4 EXPECTED FUTURE SPOT PRICE[4]

Consider a commodity with spot price P_0 and forward price F_0 at the time $t = 0$ (today). The forward contract matures in T years. If an investor takes a long position in the forward contract, they will pay F_0 at time T to purchase the commodity. One might take this position if they expect the spot price at time T, denoted P_T, to be higher than the contractual forward rate F_0.

Since forward contracts require no initial outlay, the investor would instead invest an amount equal to the present value of the forward price, $F_0e^{-r_fT}$, into an account earning the risk-free rate r_f. This way, when the

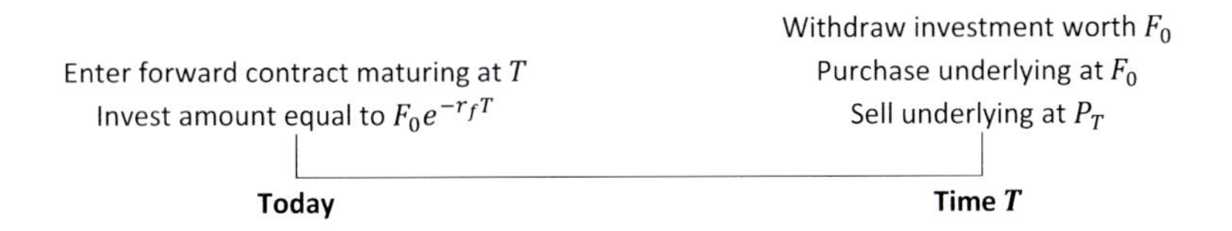

FIGURE 10.1 Forward contract mechanics.

contract matures at time T, this investment will be worth $F_0 e^{-r_f T} e^{r_f T} = F_0$, the exact amount needed to purchase the underlying in the forward contract.

After purchasing the underlying for F_0 at maturity, the investor would sell the underlying for P_T, earning profit on the difference between these two. Figure 10.1 lays out these steps more clearly.

Returning to today, we have an actual cash outflow and an expected future cash inflow. The amount of the cash outflow is $F_0 e^{-r_f T}$. The present value of the expected future cash inflow is $E(P_T)e^{-rT}$. Notice that the expected future cash inflow is discounted at the opportunity cost of capital r.

Putting these cash flows together and noting that the cash outflow is represented with a negative sign, the investor can expect to earn:

$$-F_0 e^{-r_f T} + E(P_T)e^{-rT}$$

This represents the expected profit for an investor taking a long position in a forward contract. At the initiation of a derivative contract, the contract is priced such that the expected profit is zero (otherwise an up-front payment must be made from one party to another). With the zero-profit assumption at contract initiation, we have

$$-F_0 e^{-r_f T} + E(P_T)e^{-rT} = 0$$

$$F_0 e^{-r_f T} = E(P_T)e^{-rT}$$

$$F_0 = E(P_T)\frac{e^{-rT}}{e^{-r_f T}}$$

$$= E(P_T)e^{-rT} e^{r_f T}$$

$$= E(P_T)e^{(r_f - r)T}$$

The important relation here is that of r_f and r. As discussed in Section 6.2.4, the opportunity cost of capital r might be thought of as a function of the systemic risk of the underlying commodity. If the correlation between the returns of the underlying commodity and the returns of the broad equity market is zero, then no systemic risk is present, meaning $r = r_f$. In this case, we have

$$F_0 = E(P_T)$$

In words, the forward rate is an unbiased estimate of the expected future spot rate. In the presence of systemic risk (or negative systemic risk), r will be greater than (less than) r_f, meaning the forward price will be below (above) the expected future spot price.

NOTES

1. Polynomial functions are those consisting of a sum of a variable raised to different powers multiplied by coefficients (e.g., $ax^4 + bx^3 + cx^2 + dx + e$). Quadratic functions are polynomial functions where the largest power is two (e.g., $ax^2 + bx + c$). Roots of a polynomial function are the values of the variable that equate the function to zero (e.g., $x : ax^2 + bx + c = 0$).
2. Spot rates are often proxied with zero rates. See Endnote 1 in Section 3, for a refresher on zero rates.
3. This relationship must hold true or else an arbitrage opportunity would be present.
4. This section is an adaptation of Hull (2018).

References

de Azpilcueta, Martin. "Commentary on the Resolution of Money." *Journal of Markets & Morality*, Translated by Jeannine Emery and Rodrigo Munoz, vol. 7, no. 1, 2004, pp. 171–312.

Black, Fischer. "Capital Market Equilibrium with Restricted Borrowing." *The Journal of Business*, vol. 45, no. 3, 1972, p. 444, https://doi.org/10.1086/295472.

Black, Fischer, and Myron Scholes. "The Pricing of Options and Corporate Liabilities." *Journal of Political Economy*, vol. 81, no. 3, 1973, pp. 637–654, https://doi.org/10.1086/260062.

Boardman, Anthony E., et al. *Cost-Benefit Analysis: Concepts and Practice*, 4th ed. Prentice Hall, Hoboken, NJ, 2010.

Brackenborough, Susie, et al. "The Emergence of Discounted Cash Flow Analysis in the Tyneside Coal Industry C.1700-1820." *The British Accounting Review*, vol. 33, no. 2, 2001, pp. 137–155, https://doi.org/10.1006/bare.2001.0158.

Caranti, Pedro J. "Martín de Azpilcueta: The Spanish Scholastic on Usury and Time-Preference" *Studia Humana*, vol. 9, no. 2, 2020, pp. 28–36. https://doi.org/10.2478/sh-2020-0010.

Cornuejols, Gerard, et al. *Optimization Methods in Finance*. Cambridge University Press, Cambridge, 2018.

Cox, John C., et al. "Option Pricing: A Simplified Approach." *Journal of Financial Economics*, vol. 7, no. 3, 1979, pp. 229–263, https://doi.org/10.1016/0304-405x(79)90015-1.

Fama, Eugene F., and Kenneth R. French. "The Capital Asset Pricing Model: Theory and Evidence." *Journal of Economic Perspectives*, vol. 18, no. 3, 2004, pp. 25–46, https://doi.org/10.1257/0895330042162430.

Fisher, Irving. *The Theory of Interest as Determined by Impatience to Spend Income and Opportunity to Invest It*. Macmillan, New York, 1930.

Gordon, Myron J., and Eli Shapiro. "Capital Equipment Analysis: The Required Rate of Profit." *Management Science*, vol. 3, no. 1, 1956, pp. 102–110.

Gupta, Aparna. *Risk Management and Simulation*. CRC Press/Taylor & Francis Group, Boca Raton, FL, 2014.

Hull, John. *Options, Futures, and Other Derivatives*. Pearson, London, 2018.

Jensen, Johan L. "Sur Les Fonctions Convexes Et Les Inégalités Entre Les Valeurs Moyennes." *Acta Mathematica*, vol. 30, 1906, pp. 175–193, https://doi.org/10.1007/bf02418571.

Knight, Frank H. *Risk, Uncertainty and Profit*. Houghton Mifflin, Boston, MA, 1948.

Lintner, John. "Security Prices, Risk, and Maximal Gains from Diversification." *The Journal of Finance*, vol. 20, no. 4, 1965a, p. 587, https://doi.org/10.2307/2977249.

Lintner, John. "The Valuation of Risk Assets and the Selection of Risky Investments in Stock Portfolios and Capital Budgets." *The Review of Economics and Statistics*, vol. 47, no. 1, 1965b, p. 13, https://doi.org/10.2307/1924119.

Lo, Andrew, and Archie Craig MacKinlay. *Stock Market Prices Do Not Follow Random Walks: Evidence from a Simple Specification Test*. National Bureau of Economic Research, Cambridge, MA, 1987, https://doi.org/10.3386/w2168.

Macaulay, Frederick R. *Some Theoretical Problems Suggested by the Movements of Interest Rates, Bond Yields and Stock Prices in the United States since 1856*. NBER, Cambridge, MA, 1938.

Markowitz, Harry. "Portfolio Selection." *The Journal of Finance*, vol. 7, no. 1, 1952, p. 77, https://doi.org/10.2307/2975974.

Merton, Robert C. "On the Pricing of Corporate Debt: The Risk Structure of Interest Rates." *The Journal of Finance*, vol. 29, no. 2, 1974, p. 449, https://doi.org/10.2307/2978814.

Modigliani, Franco, and Merton H. Miller. "The Cost of Capital, Corporation Finance, and the Theory of Investment: Reply." *The American Economic Review*, vol. 55, no. 3, 1965, pp. 524–27, JSTOR, https://www.jstor.org/stable/1814566.

Mossin, Jan. "Equilibrium in a Capital Asset Market." *Econometrica*, vol. 34, no. 4, 1966, p. 768, https://doi.org/10.2307/1910098.

Noonan, John T. *The Scholastic Analysis of Usury*. Harvard University Press, Cambridge, MA, 1957.

Ross, Sheldon Mark. *Introduction to Probability Models*. Elsevier, Amsterdam, Netherlands, 2019.

Sharpe, William F. "Capital Asset Prices: A Theory of Market Equilibrium under Conditions of Risk." *The Journal of Finance*, vol. 19, no. 3, 1964, p. 425, https://doi.org/10.2307/2977928.

Tobin, James. "Liquidity Preference as Behavior towards Risk." *The Review of Economic Studies*, vol. 25, no. 2, 1958, p. 65, https://doi.org/10.2307/2296205.

Treynor, Jack L. "Market Value, Time, and Risk." *SSRN Electronic Journal*, 1961, https://doi.org/10.2139/ssrn.2600356.

Tsay, Ruey S. *Analysis of Financial Time Series*. Wiley, Hoboken, NJ, 2010.

Wiener, Norbert. "Differential-Space." *Journal of Mathematics and Physics*, vol. 2, no. 1–4, 1923, pp. 131–174, https://doi.org/10.1002/sapm192321131.

Wooldridge, Jeffrey M. *Introductory Econometrics: A Modern Approach*. Cengage Learning, Boston, MA, 2018.

Index